Advanced Maths Essentials
Core 3 for OCR

Welcome to Advanced Maths Essentials: Core 3 for OCR. This book will help you to improve your examination performance by focusing on the essential maths skills you will need in your OCR Core 3 examination. It has been divided by chapter into the main topics that need to be studied. Each chapter has then been divided by sub-headings, and the description below each sub-heading gives the OCR specification for that aspect of the topic.

The book contains scores of worked examples, each with clearly set-out steps to help solve the problem. You can then apply the steps to solve the Skills Check questions in the book and past exam questions at the end of each chapter. If you feel you need extra practice on any topic, you can try the Skills Check Extra exercises on the accompanying CD-ROM. At the back of this book there is a sample exam-style paper to help you test yourself before the big day.

Some of the questions in the book have a ⊚ symbol next to them. These questions have a PowerPoint® solution (on the CD-ROM) that guides you through suggested steps in solving the problem and setting out your answer clearly.

Using the CD-ROM

To use the accompanying CD-ROM simply put the disc in your CD-ROM drive, and the menu should appear automatically. If it doesn't automatically run on your PC:

1. Select the My Computer icon on your desktop.
2. Select the CD-ROM drive icon.
3. Select Open.
4. Select core3_for _ocr.exe.

If you don't have PowerPoint® on your computer you can download PowerPoint 2003 Viewer®. This will allow you to view and print the presentations. Download the viewer from http://www.microsoft.com

Pearson Education Limited
Edinburgh Gate
Harlow
Essex
CM20 2JE
England
www.longman.co.uk

First published 2005
10 9
ISBN 978-0-582-83655-6

Design by Ken Vail Graphic Design

Cover design by Raven Design

Typeset by Tech-Set, Gateshead

Printed in China (CTPS/09)

The publisher wishes to draw attention to the Single-User Licence Agreement at the back of the book. Please read this agreement carefully before installing and using the CD-ROM.

The Publisher and Authors would like to thank Rosemary Smith for her significant contributions to Chapters 1 and 4 of this book.

We are grateful for permission from OCR to reproduce past exam questions. All such questions have a reference in the margin. OCR can accept no responsibility whatsoever for accuracy of any solutions or answers to these questions.

Every effort has been made to ensure that the structure and level of sample question papers matches the current specification requirements and that solutions are accurate. However, the publisher can accept no responsibility whatsoever for accuracy of any solutions or answers to these questions. Any such solutions or answers may not necessarily constitute all possible solutions.

1 Algebra and functions

1.1 Functions: domain and range

Understand the terms function, domain, range, one-one function. Identify the range of a given function in simple cases. Determine whether or not a function is one-one.

First consider a **mapping**, which is a relationship between two sets of data. The set of elements being mapped is called the **domain** and the resulting set is called the **range**.

> **Note:**
> A mapping can be described in words or using algebra and may be represented by a graph.

For example, consider the domain, consisting of the set {4, 3, 2, 1}, being mapped to the range, consisting of the set {8, 6, 4, 2}.

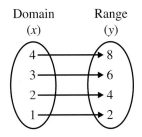

If the elements in the domain are represented by x and the elements in the range by y, the relationship is $y = 2x$.

There are four types of mapping:

> **Note:**
> The expression one-one is read as one-to-one.

- **one-one mapping**, where each element in the domain is mapped to just one element in the range;

- **many-one mapping**, where more than one element in the domain can be mapped to an element in the range;

- **one-many mapping**, where an element in the domain can be mapped to more than one element in the range;

- **many-many mapping**, where more than one element in the domain can be mapped to more than one element in the range.

A **function** is a special mapping which satisfies both these conditions:

> **Note:**
> Mappings that are one-many or many-many are **not** functions.

- it is defined for all elements of the domain;

- it is either **one-one** or **many-one**.

Here are two examples of **one-one functions**:

> **Note:**
> $x \in \mathbb{R}$ means that x is any real number.

$$f : x \mapsto 3 - x, x \in \{3, 4, 5, 6\} \qquad f(x) = 2x - 1, x \in \mathbb{R}$$

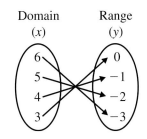

Range is {0, −1, −2, −3}

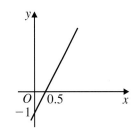

> **Note:**
> Each element in the domain maps to exactly one element in the range.

Range is f$(x) \in \mathbb{R}$

Here are two examples of **many-one functions**

$$f(x) = x^4, x \in \{-2, -1, 0, 1, 2\}$$

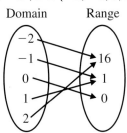

Domain   Range

Range is {0, 1, 16}

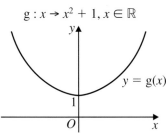

$$g : x \to x^2 + 1, x \in \mathbb{R}$$

Range is $g(x) \in \mathbb{R}$, $g(x) \geqslant 1$

Note:
More than one element in the domain can map to an element in the range.

When a function is drawn as a graph it is easy to see the domain and range as the domain is the set of all possible x-values and the range is the set of all possible y-values.

Example 1.1 Each diagram shows a sketch of the given mapping $x \mapsto y$. State, with a reason, whether the mapping is a function. If it is a function, state its range.

a $x^2 + y^2 = 25, x \in \mathbb{R}$

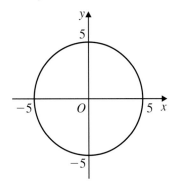

b $y = 4 - x^2, x \in \mathbb{R}$

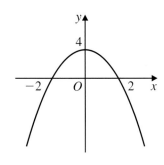

Step 1: Check the conditions for a mapping to be a function and make your conclusion.

a The mapping is many-many, since, for example, when $x = 3$, $y = \pm 4$ and also when $x = -3$, $y = \pm 4$.

Also the mapping is not defined for all elements of the domain, for example when $x = 6$, $y^2 = 25 - 36 = -9$, which has no real solutions.

Hence $x^2 + y^2 = 25$, $x \in \mathbb{R}$, is not a function.

Recall:
Many-many mappings are not functions.

Step 2: If it is a function, use the sketch to state its range.

b The mapping is many-one and it is defined for all elements of the domain. So $y = 4 - x^2$, $x \in \mathbb{R}$, is a function.

The range is $y \in \mathbb{R}$, $y \leqslant 4$.

Tip:
Be careful with the use of $<$ and $\leqslant$. In this example, when $x = 0$, $y = 4$, so the range is $y \leqslant 4$.

Example 1.2 The function f is such that $f(x) = x^2 - 4x - 5, x \in \mathbb{R}$.

a The diagram shows a sketch of $y = f(x)$.

  i Find the coordinates of A, B and C.

  ii Write $f(x)$ in the form $(x - a)^2 + b$ and hence, or otherwise, find the coordinates of the minimum point D.

b State the range of f.

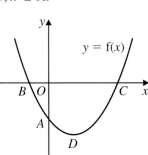

a i Consider $y = x^2 - 4x - 5$:

When $x = 0$, $y = -5$

When $y = 0$, $x^2 - 4x - 5 = 0$

$$\Rightarrow \quad (x - 5)(x + 1) = 0$$

$$\Rightarrow \quad x = 5, x = -1$$

Therefore A is the point $(0, -5)$, B is the point $(-1, 0)$ and C is the point $(5, 0)$.

ii $f(x) = x^2 - 4x - 5$

$$= (x - 2)^2 - 4 - 5$$

$$= (x - 2)^2 - 9$$

So D has coordinates $(2, -9)$.

b Range of f is $f(x) \in \mathbb{R}$, $f(x) \geqslant -9$.

Tip:
You will not get marks for drawing the graph and reading the points from it.

Recall:
$y = (x - a)^2 + b$ has minimum point at (a, b) (C1 Section 2.5).

Note:
You could use differentiation to find the coordinates of D (C1 Section 4.6).

Tip:
Look at the graph to see the set of all values that y can take. This is the range.

1.2 Composition of functions

Understand the term composition of functions and find the composition of two given functions.

Two or more functions may be combined to form a **composite function**. In doing this, you are finding the **composition** of two functions.

The notation fg is used to denote the composite function defined by $f[g(x)]$. For the composite function fg to exist, the range of g must be a subset of the domain of f.

Note:
fg means that you apply g then you apply f to the result.

Example 1.3 The functions f and g are defined as follows:

$$f: x \mapsto 2x + 3, x \in \mathbb{R} \qquad g: x \mapsto x^2 - 2, x \in \mathbb{R}$$

a Find **i** fg(x) **ii** gf(5) **iii** ff(x)

b Solve gf$(x) = -1$.

a i fg$(x) = f(x^2 - 2)$

$$= 2(x^2 - 2) + 3$$

$$= 2x^2 - 1$$

ii gf$(5) = g(2 \times 5 + 3)$

$$= g(13)$$

$$= 13^2 - 2$$

$$= 167$$

iii ff$(x) = f(2x + 3)$

$$= 2(2x + 3) + 3$$

$$= 4x + 9$$

b gf$(x) = g(2x + 3)$

$$= (2x + 3)^2 - 2$$

$$= 4x^2 + 12x + 9 - 2$$

$$= 4x^2 + 12x + 7$$

$$gf(x) = -1$$

$$\Rightarrow \quad 4x^2 + 12x + 7 = -1$$

$$4x^2 + 12x + 8 = 0$$

$$x^2 + 3x + 2 = 0$$

$$(x + 1)(x + 2) = 0$$

$$x = -1, x = -2$$

Tip:
fg$(x) \neq$ f$(x) \times$ g(x)

Note:
ff(x) is sometimes written f$^2(x)$.

Tip:
ff$(x) \neq$ f$(x) \times$ f(x)

Note:
In general fg$(x) \neq$ gf(x).

Note:
This doesn't mean find gf(-1).

Tip:
If the quadratic doesn't factorise, use the quadratic formula.

Example 1.4 The functions f and g are defined as follows:

$$\text{f}: x \mapsto x^2, x \in \mathbb{R} \qquad \text{g}: x \mapsto 3x - 2, x \in \mathbb{R}$$

a Find fg(x) and state the range of fg.

b Find gf(x) and state the range of gf.

c Find the value of *a* for which fg(*a*) = gf(*a*).

Step 1: Apply g, then apply f to the result.

a fg(x) = f(3x − 2)

$$= (3x - 2)^2$$

Step 2: State the range of fg.

Since $(3x - 2)^2 \geq 0$, the range of fg is fg(x) $\in \mathbb{R}$, fg(x) ≥ 0

Step 3: Apply f, then apply g to the result.

b gf(x) = g(x^2)

$$= 3x^2 - 2$$

Step 4: State the range of gf.

Since $3x^2 \geq 0$ for all x, $3x^2 - 2 \geq -2$, so the range of gf is gf(x) $\in \mathbb{R}$, gf(x) ≥ -2

Step 5: Substitute the given variable, using your answers from **a** and **b**.

c fg(*a*) = $(3a - 2)^2$

$$= 9a^2 - 12a + 4$$

gf(*a*) = 3*a* − 2

Step 6: Set up an equation and solve for *a*.

$$\text{fg}(a) = \text{gf}(a)$$

$$\Rightarrow \qquad 9a^2 - 12a + 4 = 3a^2 - 2$$

$$6a^2 - 12a + 6 = 0$$

$$(\div 6) \qquad a^2 - 2a + 1 = 0$$

$$(a - 1)(a - 1) = 0$$

$$a = 1$$

Tip:
In **a**, use the fact that this is in completed square form so you know its minimum. (C1 Section 2.5). If you are unsure, do a sketch.

a   **b**

Tip:
In **b**, again a sketch may help.

Tip:
You could substitute your answer into fg and gf to check that you get the same result.

1.3 Inverse functions

Find the inverse of a one-one function in simple cases. Illustrate in graphical terms the relation between a one-one function and its inverse.

The **inverse function**, f^{-1}, of a function f, represents the reverse mapping. You must learn the following:

- For a function to have an inverse, it must be one-one.

- The domain of f^{-1} is the range of f.

- The range of f^{-1} is the domain of f.

For example the function f: $x \mapsto \frac{1}{2}x + 1$ with domain {3, 2, 1, 0}, has range {2.5, 2, 1.5, 1}. The inverse function sends each element from the range of f back to its original value in the domain of f, so under f^{-1}: 2.5 $\mapsto$ 3, 2 $\mapsto$ 2, 1.5 $\mapsto$ 1 and 1 $\mapsto$ 0.

Note:
The notation $\text{f}^{-1}(x)$ is easy to confuse with $\text{f}'(x)$ and $(\text{f}(x))^{-1}$. Remember that $\text{f}^{-1}(x)$ is the inverse function, $\text{f}'(x)$ is the derivative and $(\text{f}(x))^{-1}$ is the reciprocal function $\dfrac{1}{\text{f}(x)}$.

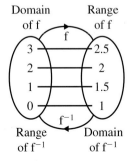

The domain of f^{-1} is {2.5, 2, 1.5, 1} and the range of f^{-1} is {3, 2, 1, 0}.

Now extend the domain so that f: $x \mapsto \frac{1}{2}x + 1$ for $x \in \mathbb{R}$.

We can illustrate the function on a graph by the line $y = \frac{1}{2}x + 1$.

To show the inverse function, use the fact that the graphs of a function and its inverse are reflections of each other in the line $y = x$.

Tip:

To show this on a diagram, ensure that the scales on both axes are the same.

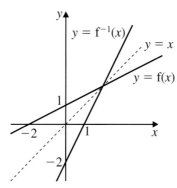

Note:

In this example, the inverse function can be illustrated by a line.

So if (x, y) lies on f(x), then (y, x) lies on f$^{-1}(x)$.

To find the inverse function f^{-1}, let

$$y = \frac{1}{2}x + 1$$

Then interchange x and y:

$$x = \frac{1}{2}y + 1$$

Note:

You can interchange x and y at the beginning of the working or at the end.

Now make y the subject:

$$\frac{1}{2}y = x - 1$$
$$y = 2x - 2$$

Note:

This is the equation of the line $y = f^{-1}(x)$ in the diagram above.

Write this as a function:

So f^{-1}: $x \mapsto 2x - 2, x \in \mathbb{R}$

Note:

Always write f$^{-1}(x)$ as a function of x.

If a function is many-one, by restricting its domain it can be made into a one-one function, enabling an inverse function to be found.

For example, the function f: $x \mapsto x^2, x \in \mathbb{R}$ is a many-one function since, for example, f(-3) $= (-3)^2 = 9$ and f(3) $= 3^2 = 9$.

If the domain is restricted to $x \geq 0$, then f is a one-one function and its inverse f^{-1} can be found, where f$^{-1}(x) = \sqrt{x}$.

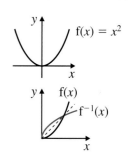

Example 1.5 The function f has domain $3 \leq x \leq 9$ and is defined by $f(x) = \dfrac{8}{1 - x} + 6$. A sketch of $y = f(x)$ is shown below.

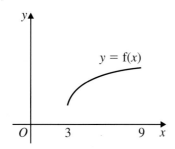

a Calculate f(3) and f(9).

b Find the range of f.

c The inverse function is f^{-1}. Find f^{-1}, stating its domain and range.

d On the same set of axes, sketch $y = f(x)$ and $y = f^{-1}(x)$.

Step 1: Substitute appropriate *x*-values into the expression for f.

Step 2: Use the values obtained in **a** and check the sketch.

Step 3: Let $y = f(x)$ then interchange *x* and *y*.

Step 4: Make *y* the subject.

Step 5: Use the relationship between f and f^{-1} to state the domain and range.

Step 6: Reflect the given curve in the line $y = x$ for the appropriate domain and range.

a $f(3) = \dfrac{8}{1-3} + 6 = 2$

   $f(9) = \dfrac{8}{1-9} + 6 = 5$

b The range of f is $2 \leqslant f(x) \leqslant 5$.

Note:
You also need to check the sketch to make sure there are no turning points.

c Let $y = \dfrac{8}{1-x} + 6$.

   Interchanging *x* and *y* gives

   $$x = \frac{8}{1-y} + 6$$

   $$x - 6 = \frac{8}{1-y}$$

   $$1 - y = \frac{8}{x-6}$$

   $$y = 1 - \frac{8}{x-6}$$

   So $f^{-1}(x) = 1 - \dfrac{8}{x-6}$.

   The domain of f^{-1} is $2 \leqslant x \leqslant 5$ and the range is $3 \leqslant f^{-1}(x) \leqslant 9$.

d

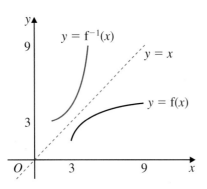

In the above example you can check your inverse function by substituting a value for *x* in the original function, for example:

$$f(3) = \frac{8}{1-3} + 6 = 2$$

Now substitute the result into the inverse function.

Note:
f maps 3 to 2 and f^{-1} maps 2 back to 3.

$$f^{-1}(2) = 1 - \frac{8}{2-6} = 3$$

So, because the inverse function reverses the process of the original function

$$f^{-1}f(x) = x$$

It is also true that $ff^{-1}(x) = x$

This property can be used to solve equations.

Example 1.6 The function $f(x)$ is defined by

$$f(x) = \sqrt{x + 2},\ x \in \mathbb{R},\ x \geqslant -2.$$

a Find an expression for f^{-1}, stating the domain and range of f^{-1}.

b Find the value of x for which $f(x) = \frac{1}{2}$.

c On the same set of axes, sketch $y = f(x)$ and $y = f^{-1}(x)$.

d Solve $f(x) = f^{-1}(x)$.

Step 1: Let $y = f(x)$ then interchange x and y.

a Let $y = \sqrt{x + 2}$

Interchanging x and y gives

$x = \sqrt{y + 2}$

Step 2: Make y the subject

$x^2 = y + 2$

$y = x^2 - 2$

$f^{-1}(x) = x^2 - 2$

> **Tip:**
> Square both sides.

Step 3: Use the relationship between f and f^{-1} to state the domain and range.

Since the range of $f(x)$ is $f(x) \in \mathbb{R}$, $f(x) \geqslant 0$, the domain of f^{-1} is $x \in \mathbb{R}$, $x \geqslant 0$.

Since the domain of f is $x \in \mathbb{R}$, $x \geqslant -2$, the range of f^{-1} is $f^{-1}(x) \in \mathbb{R}$, $f^{-1}(x) \geqslant -2$.

> **Tip:**
> Make it clear for which function you are stating the domain or range.

Step 4: Use the inverse function as the reverse process.

b $f(x) = \frac{1}{2}$

$\Rightarrow\ x = f^{-1}\left(\frac{1}{2}\right)$

$= \left(\frac{1}{2}\right)^2 - 2 = -\frac{7}{4}$

So $x = -\frac{7}{4}$.

> **Tip:**
> Often candidates set up the equation $f(x) = \frac{1}{2} \Rightarrow \sqrt{x + 2} = \frac{1}{2}$ and solve it. That is an acceptable alternative method, but it may take longer.

Step 5: Sketch $y = f(x)$ for the domain and reflect it in $y = x$.

c

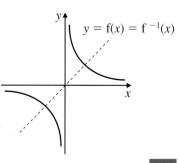

Step 6: Set up an equation in x and solve.

d From the graph, $f(x) = f^{-1}(x)$ at the point of intersection with the line $y = x$, i.e. when $f(x) = x$ and $f^{-1}(x) = x$.

Using $f^{-1}(x) = x$ gives

$x^2 - 2 = x$

$x^2 - x - 2 = 0$

$(x - 2)(x + 1) = 0$

$x = 2$ or -1

But $x \geqslant 0$, so $x = 2$.

> **Tip:**
> It is much easier to solve $f(x) = x$ or $f^{-1}(x) = x$, rather than $f(x) = f^{-1}(x)$ for which you would have to solve $\sqrt{x + 2} = x^2 - 2$.
> Here $f^{-1}(x) = x$ is the easiest method as there are no $\sqrt{\ }$ signs to deal with.

> **Note:**
> From the sketch you can see that the intersection occurs when $x \geqslant 0$.

In the above example, the equation $f(x) = f^{-1}(x)$ is solved by finding any points at which $f(x) = x$ or $f^{-1}(x) = x$.

In general, $f(x) = f^{-1}(x)$ for all elements in the domain, the function f is **self inverse** and f^{-1} is the same as f. In this case, $ff^{-1}(x) = x$, $f^{-1}f(x) = x$ and $ff(x) = x$. The line $y = x$ is a line of symmetry of the graph of $y = f(x)$.

An example of a self-inverse function is f, defined by $f(x) = \dfrac{1}{x}$, $x \neq 0$, since $f^{-1}(x) = \dfrac{1}{x}$, $x \neq 0$.

1 State whether each of the following mappings represents a function.
If it is a function, state whether it is one-one or many-one.

a $y = 2(5 - x), x \in \mathbb{R}$

b $y^2 = 1 - x^2, x \in \mathbb{R}$

c $y = \dfrac{1}{x}, x \in \mathbb{R}, x \neq 0$

d $y = x^2, x \in \mathbb{R}$

2 Find the range of each of the following functions:

a $f(x) = \dfrac{x + 5}{2}, x = \{0, 1, 2, 3\}$

b $f: x \mapsto \dfrac{1}{5 - x}, x = \{1, 2, 3, 4\}$

c $g: x \mapsto x^3, x \in \mathbb{R}, x \geqslant 0$

d $f(x) = x^2 - 2, x \in \mathbb{R}$

3 For each of the following functions, sketch the function and state its range.

a $f: x \mapsto 4x - 3, x \in \mathbb{R}$

 b $g(x) = \sin x°, x \in \mathbb{R}, 0 \leqslant x \leqslant 360$

c $f(x) = \dfrac{1}{x^2}, x \in \mathbb{R}, x \neq 0$

d $h: x \mapsto x^2 - 6x, x \in \mathbb{R}$

 4 The function f is defined by $f: x \mapsto \dfrac{6}{x} + 2x$ for the domain $1 \leqslant x \leqslant 3$. Find the range of f.

5 Functions f, g and h are defined as follows:

$f: x \mapsto 1 - 2x, x \in \mathbb{R}$ $g: x \mapsto x^2 + 3, x \in \mathbb{R}$ $h: x \mapsto \dfrac{x + 5}{2}, x \in \mathbb{R}$

Find the value of

a $fg(2)$

b $hf(-1)$

c $gg(0)$

d $fh(-11)$

e $gf(\tfrac{1}{4})$

f $hgf(1.5)$

6 Functions f, g and h are defined as follows:

$f: x \mapsto 2^x, x \in \mathbb{R}$ $g: x \mapsto 3x + 2, x \in \mathbb{R}$ $h: x \mapsto \dfrac{1}{x}, x \in \mathbb{R}, x \neq 0$

Find the composite functions

a fg   **b** hh   **c** gh   **d** hg

7 Functions f, g and h are defined as follows:

$f(x) = 2x + 9, x \in \mathbb{R}$ $g(x) = \log_{10} x, x \in \mathbb{R}, x > 0$ $h(x) = 1 - x^2, x \in \mathbb{R}$

Solve these equations:

a $ff(x) = 9$ **b** $gf(x) = 0$ **c** $fh(x) = -5$ **d** $hf(x) = -8$

8 Which of the following functions has an inverse?

a

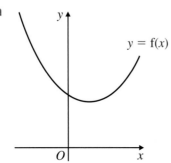

$y = f(x)$

b

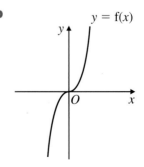

$y = f(x)$

9 For each of the following functions, f,

   i find an expression for $f^{-1}(x)$ and state the domain and range of f^{-1},

   ii on the same set of axes, sketch $y = f(x)$ and $y = f^{-1}(x)$.

  a $f: x \mapsto 2x + 5, x \in \mathbb{R}$ **b** $f: x \mapsto \dfrac{3 - x}{4}, x \in \mathbb{R}$

  c $f: x \mapsto x^2, x \in \mathbb{R}, x \geqslant 0$ **d** $f: x \mapsto \sqrt{x - 3}, x \in \mathbb{R}, 3 \leqslant x \leqslant 12$

10 Functions f, g and h are defined as follows:

  $f(x) = 5x - 4, x \in \mathbb{R}$ $g(x) = 1 - 2x, x \in \mathbb{R}$ $h(x) = x^2, x \in \mathbb{R}$

  Solve these equations:

  a $gf(x) = g^{-1}(x)$ **b** $h(x) = g^{-1}(x)$ **c** $hg(x) = h(x)$

11 A function is defined by $f: x \mapsto 3 - \dfrac{2}{x}, x \in \mathbb{R}, x \neq 0$.

  a Find f^{-1} and state the value of x for which f^{-1} is undefined.

  b Find the values of x for which $f(x) = f^{-1}(x)$.

12 A function is defined by $f: x \mapsto \dfrac{1}{5 - 4x}, x \in \mathbb{R}, x \neq \frac{5}{4}$.

  a Find the values of x which map onto themselves under the function f.

  b Find an expression for f^{-1}.

  Another function is defined by $g: x \mapsto x^2 - 3$.

  c Evaluate $gf(1)$.

SKILLS CHECK   **1A EXTRA** is on the CD

1.4 Compositions of transformations

Use and recognise compositions of transformations of graphs, such as the relationships
between the graphs of $y = f(x)$ and $y = af(x + b)$.

In *Core 1* you learnt that transformations applied to $y = f(x)$ have the
following effects:

Translations

$y = f(x) + a$ represents a translation by a units in the y-direction.

$y = f(x + a)$ represents a translation by $-a$ units in the x-direction.

Stretches

$y = af(x)$ represents a stretch by scale factor a in the y-direction.

$y = f(ax)$ represents a stretch by scale factor $\frac{1}{a}$ in the x-direction.

Reflections

$y = -f(x)$ represents a reflection in the x-axis.

$y = f(-x)$ represents a reflection in the y-axis.

Recall:
Transformations (C1 Section 3.8).

Tip:
The vector form of a translation by a units in the y-direction is $\begin{pmatrix} 0 \\ a \end{pmatrix}$.

Tip:
The vector form of a translation by $-a$ units in the x-direction is $\begin{pmatrix} -a \\ 0 \end{pmatrix}$.

Compositions of transformations

In *Core 3* you need to be able to apply **compositions** of transformations.
In some cases, the order in which the transformations are applied is
important.

Here are some examples:

$y = 2f(5x)$ is a stretch of $y = f(x)$ by scale factor $\frac{1}{5}$ in the x-direction and a stretch by scale factor 2 in the y-direction. If, for example, $f(x) = \sin x$, then $y = 2 \sin 5x$.

Note:
$y = f(x)$ is mapped to
$y = af(kx)$.

$y = -f(x) + 2$ is a reflection of $y = f(x)$ in the x-axis and a translation by 2 units in the y-direction. The vector of the translation is $\begin{pmatrix} 0 \\ 2 \end{pmatrix}$. If, for example, $f(x) = x^3$ then $y = -x^3 + 2$.

Note:
$y = f(x)$ is mapped to
$y = b - f(x)$.

$y = f(x - 4) + 3$ is a translation of $y = f(x)$ by 4 units in the x-direction and 3 units in the y-direction. The vector of this translation is $\begin{pmatrix} 4 \\ 3 \end{pmatrix}$. If, for example, $f(x) = x^2$ then $y = (x - 4)^2 + 3$.

Note:
$y = f(x)$ is mapped to
$y = f(x - a) + b$.
You used this in C1 to sketch quadratic curves (C1 Sections 2.5, 3.8).

$y = 3f(x + 1)$ is a stretch of $y = f(x)$ by scale factor 3 in the y-direction and a translation by -1 unit in the x-direction. If, for example, $f(x) = x^2$ then $y = 3(x + 1)^2$.

Note:
$y = f(x)$ is mapped to
$y = af(x + b)$.

$y = 2f(x) + 5$ is a stretch by scale factor 2 in the y-direction followed by a translation 5 units in the y-direction. If, for example, $f(x) = x^2$, then $y = 2x^2 + 5$. In this example the order in which the transformations are carried out is important. If the translation is applied first, then $y = 2(x^2 + 5)$.

Note:
$y = f(x)$ is mapped to
$y = af(x) + b$.

Sketching curves

Compositions of transformations can be used to sketch curves.

Example 1.7 The diagram shows a sketch of $y = f(x)$.

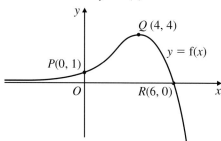

In each of the following, sketch the curve, showing the coordinates of P_1, Q_1, R_1, the images of P, Q and R:

a $y = 4 - f(x)$ **b** $y = \frac{1}{2}f(x + 3)$ **c** $y = \frac{1}{4}f(2x)$

Step 1: Apply the transformations to the given curve.

a This is a reflection of $y = f(x)$ in the x-axis, followed by a translation of 4 units in the y-direction.

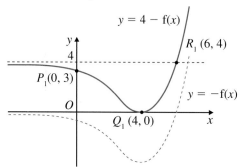

Tip:
The broken curve is the reflection. The solid curve is the final transformation. Make this clear on your sketch.

Tip:
The reflection in the x-axis makes all the y-coordinates negative and then the vertical translation adds 4 to the y-coordinates. The x-values remain unchanged.

Step 2: Find the coordinates of the images of the given points. The images of P, Q and R are $(0, 3)$, $(4, 0)$ and $(6, 4)$ respectively.

Step 3: Apply the transformations to the given curve.

b This is a translation of $y = f(x)$ by -3 units in the x-direction, and a stretch by scale factor $\frac{1}{2}$ in the y-direction.

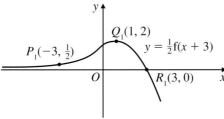

Tip:
The horizontal translation takes 3 off the x-coordinates and the vertical stretch halves the y-coordinates.

Step 4: Find the coordinates of the images of the given points.

The images of P, Q and R are $(-3, \frac{1}{2})$, $(1, 2)$ and $(3, 0)$ respectively.

Step 5: Apply the transformations to the given curve.

c This is a stretch by scale factor $\frac{1}{2}$ in the x-direction and a stretch by scale factor $\frac{1}{4}$ in the y-direction.

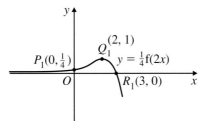

Tip:
The horizontal stretch halves the x-coordinates and the vertical stretch divides the y-coordinates by 4.

Step 6: Find the coordinates of the images of the given points.

The images of P, Q and R are $(0, \frac{1}{4})$, $(2, 1)$ and $(3, 0)$ respectively.

Example 1.8 The diagram shows a sketch of the graph $y = f(x)$, where $f: x \mapsto \sin x$, $0 \leqslant x \leqslant 2\pi$.

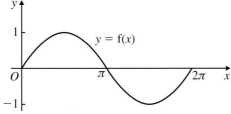

Recall:
Trigonometric graphs (C2 Section 1.5).

Given that $g: x \mapsto 3 + \sin 2x$, $0 \leqslant x \leqslant 2\pi$,

a sketch the graph of $y = g(x)$,

b state the range of g.

Step 1: Apply the transformations to the given curve.

a This is a stretch by scale factor $\frac{1}{2}$ in the x-direction, and a translation by 3 units in the y-direction.

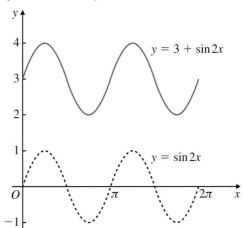

Tip:
If you draw intermediate steps, make it clear which graph is which.

Tip:
The range of $\sin x$ is $-1 \leqslant y \leqslant 1$ so, following the translation by 3 units in the y-direction, this becomes $2 \leqslant y \leqslant 4$. Notice that the horizontal stretch doesn't affect the range.

Step 2: Use the sketch to find the range.

b From the graph, the range of g is $g \in \mathbb{R}$, $2 \leqslant g(x) \leqslant 4$.

1.5 The modulus function

Understand the meaning of $|x|$ and use relations such as $|a| = |b| \Leftrightarrow a^2 = b^2$ and $|x - a| < b \Leftrightarrow a - b < x < a + b$ in the course of solving equations and inequalities. Understand the relationship between the graphs of $y = f(x)$ and $y = |f(x)|$.

The **modulus** (or absolute value or magnitude) of a number is its positive numerical value.

The modulus of x, $|x|$, is defined as

$$|x| = x, \quad x \in \mathbb{R}, \quad x \geq 0$$
$$|x| = -x, \quad x \in \mathbb{R}, \quad x < 0$$

For example, $|4| = |-4| = 4$.

> **Note:**
> The modulus is denoted by vertical lines as shown.

> **Tip:**
> Your calculator may have a modulus facility. It is often called 'abs'.

The graph of a modulus function

Consider the function f. To draw the graph of $y = |f(x)|$, you need to consider the sign of $f(x)$.

When $f(x) \geq 0$, $|f(x)| = f(x)$, so the graph of $y = |f(x)|$ is the same as the graph of $y = f(x)$.

When $f(x) < 0$, $|f(x)| = -f(x)$, so the graph of $y = |f(x)|$ is a reflection in the x-axis of the graph of $y = f(x)$.

So to draw $y = |f(x)|$ from the graph of $y = f(x)$,

- draw any part of $y = f(x)$ that is on or above the x-axis

- replace any part of $y = f(x)$ that is below the x-axis with its reflection in the x-axis.

> **Tip:**
> When $f(x) > 0$, the graph of $y = f(x)$ is above the x-axis.

> **Tip:**
> When $f(x) < 0$, the graph of $y = f(x)$ is below the x-axis.

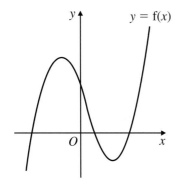

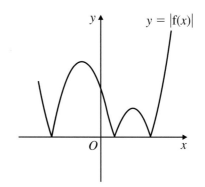

Graphical solution of equations and inequalities

Equations and inequalities involving modulus functions can be solved graphically as follows:

Example 1.9 **a** Sketch the graph of $y = 3x - 2$.

b The graph of $y = |3x - 2|$ consists of two parts. Draw a sketch to show the graph of $y = |3x - 2|$, labelling each part with its equation.

The line $y = 7$ intersects $y = |3x - 2|$ at P and Q.

c **i** Find the x-coordinates of P and Q.

ii Solve the inequality $|3x - 2| > 7$.

Step 1: Draw a sketch of
$y = 3x - 2$.

a

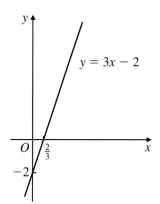

Tip:
When $x = 0$, $y = -2$ and when $y = 0$, $x = \frac{2}{3}$.

Step 2: State the equations of the lines.

Step 3: Draw a second sketch, replacing the part of the line below the x-axis with its reflection in the x-axis and label the two parts separately.

b On the graph of $y = |3x - 2|$:

- when $3x - 2 \geqslant 0$, i.e. $x \geqslant \frac{2}{3}$, the equation of the line is $y = 3x - 2$,

- when $3x - 2 < 0$, i.e. $x < \frac{2}{3}$, the equation of the line is $y = -(3x - 2)$, i.e. $y = 2 - 3x$.

Recall:
The reflection in the x-axis of $y = f(x)$ is $y = -f(x)$.

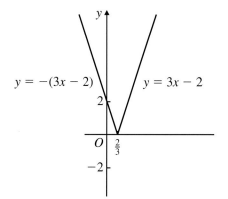

Step 4: Sketch the given line on the modulus graph, labelling the points P and Q.

c

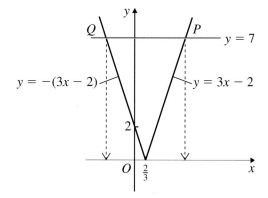

Step 5: Set up two equations and solve to find the x-coordinates of P and Q.

i At P $\qquad 3x - 2 = 7$
$$3x = 9$$
$$x = 3$$

At Q $\quad -(3x - 2) = 7$
$$3x - 2 = -7$$
$$3x = -5$$
$$x = -\frac{5}{3}$$

So the x-coordinates of P and Q are 3 and $-\frac{5}{3}$.

Note:
If $|f(x)| = c$ then $f(x) = c$ or $-f(x) = c$.

Step 6: Use the graphs to solve the inequality.

ii From the sketch:

$$|3x - 2| > 7 \text{ when } x < -\frac{5}{3} \text{ or } x > 3.$$

Tip:
Find where the graph of $y = |3x - 2|$ is above the graph of $y = 7$.

Example 1.10 Solve the inequality $|2x + 5| < |x - 3|$.

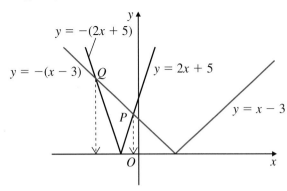

Step 1: Draw a sketch of $y = |2x + 5|$.

Step 2: On the same set of axes, draw a sketch of $y = |x - 3|$ and label the points of intersection.

Step 3: Find the points of intersection of the graphs.

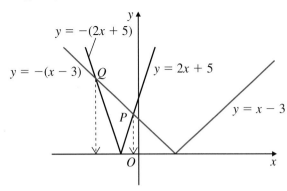

Let P be the point of intersection of $y = 2x + 5$ and $y = -(x - 3)$.

At P:
$$2x + 5 = -(x - 3)$$
$$2x + 5 = -x + 3$$
$$3x = -2$$
$$x = -\tfrac{2}{3}$$

Let Q be the point of intersection of $y = -(2x + 5)$ and $y = -(x - 3)$.

Step 4: Use the x-coordinates of the points of intersection and the graphs to solve the inequality.

At Q:
$$-(2x + 5) = -(x - 3)$$
$$2x + 5 = x - 3$$
$$x = -8$$

So $|2x + 5| < |x - 3|$ when $-8 < x < -\tfrac{2}{3}$.

Example 1.11 The diagram shows a sketch of $y = \sin x, 0° \leqslant x \leqslant 360°$.

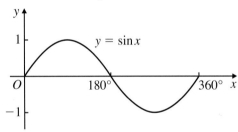

For $0° \leqslant x \leqslant 360°$,

a draw a sketch showing $y = |\sin x|$ and $y = 0.5$,

b solve the equation $|\sin x| = 0.5$.

Step 1: Replace the given curve below the x-axis with its reflection in the x-axis.

a

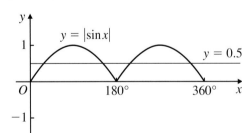

Step 2: Draw the given line on the same set of axes.

Step 3: Solve the appropriate trig equations.

b
$$|\sin x| = 0.5$$
$$\Rightarrow \quad \sin x = 0.5$$
$$x = 30°, 150°$$
$$\text{or} \quad -\sin x = 0.5$$
$$\sin x = -0.5$$
$$x = 210°, 330°$$

So $x = 30°, 150°, 210°, 330°$.

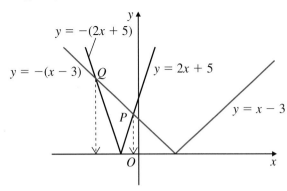

Tip:
Label the separate parts of the modulus graph with their equations. This will make it easier to see which lines intersect.

Tip:
Find the x-values for which the graph of $y = |2x + 5|$ is below the graph of $y = |x - 3|$.

Recall:
Trigonometric graphs (C2 Section 1.5).

Tip:
Notice that two points of intersection are with the original part of the curve, i.e. $y = \sin x$, and the other two are with the reflected part of the curve, i.e. $y = -\sin x$.

Recall:
Find the principal value from the calculator and then find other solutions in range (C2 Section 1.7).

Example 1.12   It is given that $f(x) = x^2 - 6x$, $x \in \mathbb{R}$.

The graph of $y = f(x)$ crosses the x-axis at the origin and B and has a minimum point at C.

a   Draw a sketch of $y = f(x)$, giving the coordinates of B and C.

b   On a separate set of axes, sketch $y = |f(x)|$ showing the coordinates of any intercepts with the axes and any stationary points.

c   State the range of y, where $y = |f(x)|$.

Step 1: Substitute $y = 0$ and $x = 0$ to find the intercepts with the axes.

a   Let $y = x^2 - 6x$

When $x = 0$, $y = 0$

When $y = 0$, $x^2 - 6x = 0$

$$x(x - 6) = 0$$

$$x = 0 \text{ or } x = 6$$

B is the point $(6, 0)$.

Step 2: Complete the square to find the minimum point.

$y = (x - 3)^2 - 9$, so the minimum point C is at $(3, -9)$.

Recall:
Quadratic curves (C1 Section 3.7).

Recall:
Completing the square (C1 Section 2.5).

Tip:
You could differentiate to find the minimum (C1 Section 4.6).

Step 3: Sketch the curve.

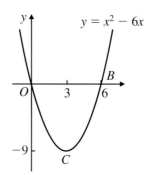

Step 4: Sketch $y = |f(x)|$ by reflecting the part below the x-axis in the x-axis.

b

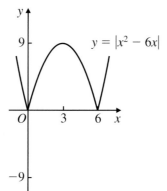

Step 5: Use symmetry to find the coordinates of the maximum point.

The graph meets the x-axis at $(0, 0)$ and $(6, 0)$.

By symmetry, the maximum point is $(3, 9)$.

Step 6: Use your sketch to state the range.

c   The range of $y = |f(x)|$ is $y \in \mathbb{R}$, $y \geq 0$.

Recall:
The range is the set of possible values for y.

Algebraic solution of equations and inequalities

Equations and inequalities involving the modulus function can also be solved algebraically.

Note that

$$|x| < k \Rightarrow -k < x < k$$

and $|x| > k \Rightarrow x < -k$ or $x > k$

It is useful to note that, in general

$$|x - a| < b$$
$$\Rightarrow \quad -b < x - a < b$$
$$\Rightarrow \quad a - b < x < a + b$$

and

$$|x - a| > b$$
$$\Rightarrow \quad x - a < -b \text{ or } x - a > b$$
$$\Rightarrow \quad x < a - b \text{ or } x > a + b$$

Note:
The inequality can be written with x 'sandwiched' in the middle.

Recall:
You need to write two separate inequalities for x.

Example 1.13 Solve the following inequality:

$$|x - 3| < 5$$

Step 1: Write the inequality as a 'sandwich' and solve.

$$|x - 3| < 5 \Rightarrow -5 < x - 3 < 5$$
$$-5 + 3 < x < 5 + 3$$
$$-2 < x < 8$$

Also note the following relationships:

$$|a| = |b| \Leftrightarrow a^2 = b^2$$
$$|a| > |b| \Leftrightarrow a^2 > b^2$$
$$|a| < |b| \Leftrightarrow a^2 < b^2.$$

Recall:
If the inequality had been $|x - 3| > 5$, then $x - 3 > 5$ or $x - 3 < -5$, giving $x > 8$ or $x < -2$.

Note:
The symbol $\Leftrightarrow$ means that the statement is true when reading from left to right and from right to left.

Example 1.14 Solve the inequality $|2x + 5| < |x - 3|$.

Step 1: Square both sides.

$$|2x + 5| < |x - 3|$$

Step 2: Expand and simplify to obtain an inequality of the form $g(x) < 0$.

$$\Rightarrow \quad (2x + 5)^2 < (x - 3)^2$$
$$4x^2 + 20x + 25 < x^2 - 6x + 9$$
$$3x^2 + 26x + 16 < 0$$

Step 3: Solve this quadratic inequality.

$$(3x + 2)(x + 8) < 0$$

So $-8 < x < -\frac{2}{3}$.

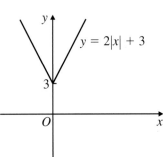

Note:
This was solved graphically in Example 1.10.

Tip:
Remember to square *both* sides.

Recall:
Quadratic inequalities (C1 Section 2.8). Sketch the curve $y = g(x)$ and find the values of x for which $g(x)$ is below the x-axis.

You may also be asked to apply **transformations** to modulus functions.

For example, consider the graph of $y = 2|x| + 3$. This is a stretch of $y = |x|$ by scale factor 2 in the y-direction to give $y = 2|x|$ followed by a translation of $y = 2|x|$ by 3 units in the y-direction.

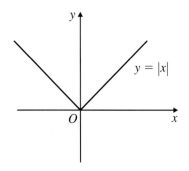

$y = |x|$

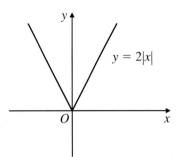

$y = 2|x|$

If $f(x) = |x|$, $y = 2f(x)$

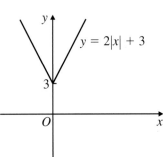

$y = 2|x| + 3$

If $g(x) = 2|x|$, $y = g(x) + 3$

1 Describe how the graph of $y = f(x)$ can be mapped to the graph of $y = g(x)$ by applying two transformations:

 a $f(x) = x$, $g(x) = \frac{1}{2}x - 3$ **b** $f(x) = 2^x$, $g(x) = 0.4(2^{-x})$ **c** $f(x) = \dfrac{1}{x}$, $g(x) = \dfrac{5}{x-2}$

2 A sketch of the graph of $y = f(x)$ is shown in the diagram. The point P has coordinates $(-2, 0)$, Q has coordinates $(0, 3)$ and R has coordinates $(2, 0)$.

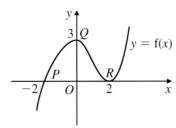

On separate sets of axes, sketch the graph of each of the following, showing the coordinates of any points at which the curve has a turning point.

 a $y = f(x - 3) + 2$ **b** $y = \frac{1}{3}f(2x)$ **c** $y = -|f(x)|$

3 a The graph of $y = \tan x$ is subjected to a stretch in the y-direction by scale factor 3 followed by a translation by -2 units in the y-direction. Write down the equation of the resulting curve.

 b The graph of $y = \sin x$ is subjected to a stretch in the x-direction by scale factor 3 followed by a reflection in the x-axis. Write down the equation of the resulting curve.

4 a Sketch the graph of $y = f(x)$, where f: $x \mapsto \cos x$, $0 \leqslant x \leqslant 2\pi$.

 Given that g: $x \mapsto -\cos\left(x + \dfrac{\pi}{4}\right)$, $0 \leqslant x \leqslant 2\pi$,

 b sketch the graph of $y = g(x)$, showing the coordinates of any turning points and intersections with the x-axis,

 c state the range of $g(x)$.

5 Sketch the graphs of the following, giving the coordinates of any points of intersection with the coordinate axes.

 a $y = |2x - 3|$ **b** $y = |1 - 2x|$ **c** $y = |x| + 4$

6 a On the same set of axes, sketch the graphs of

 $$y = |2x - 6| \quad \text{and} \quad y = |\tfrac{1}{2}x|.$$

 b Solve the equation $|2x - 6| = |\tfrac{1}{2}x|$.

 c Solve the inequality $|2x - 6| < |\tfrac{1}{2}x|$.

7 For each of the following, on separate sets of axes sketch the graph of

 i $y = f(x)$,   **ii** $y = g(x)$,

 showing the coordinates of any points at which the curve has a turning point or meets the axes.

 a $f(x) = x(x + 4)$ $g(x) = |x(x + 4)|$

 b $f(x) = \dfrac{1}{x}$ $g(x) = \dfrac{1}{|x|}$

 c $f(x) = \cos x, 0 \leqslant x \leqslant 2\pi$ $g(x) = |\cos x|, 0 \leqslant x \leqslant 2\pi$

8 Sketch the graph of each of the following functions, where a is a positive constant, giving the coordinates of any points of intersection with the coordinate axes.

    a $y = |x - a|$ **b** $y = |2x + a|$ **c** $y = |x + a| + a$

9 For each of the following:

    i on the same set of axes, sketch the graphs of $y = f(x)$ and $y = g(x)$,

    ii solve the equation $f(x) = g(x)$.

        a $f(x) = |3 - 2x|$ $g(x) = x + 1$

        b $f(x) = |3x - 5|$ $g(x) = |2x + 1|$

         c $f(x) = |x|$ $g(x) = 2|x - a|$, where a is a positive constant

 10 Solve

    a $|2x + 6| = 8$ **b** $|2x + 6| < 8$ **c** $|2x + 6| \geqslant 8$

11 Solve $|2x + 3| > |3x - 6|$.

SKILLS CHECK **1B EXTRA** is on the **CD**

1.6 The functions e^x and $\ln x$

Understand the properties of the exponential and logarithmic functions e^x and $\ln x$ and their graphs, including their relationship as inverse functions.

Understand exponential growth and decay.

The exponential function e^x

In *Core 2* you met graphs of the form $y = a^x$, known as exponential curves.

Recall that

- when $x = 0$, $y = a^0 = 1$ so the graph passes through $(0, 1)$
- for $a > 1$, as $x \to \infty$, $y \to \infty$, and, as $x \to -\infty$, $y \to 0$, so the x-axis is an asymptote.

Recall:
Exponent means index or power (C1 Section 1.1).

Recall:
An asymptote is a line that, as x tends to a particular value, the curve approaches but never meets.

Note:
The graph of a^x becomes steeper than the polynomial graphs you have studied.

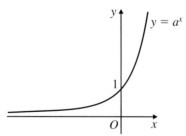

The function $f(x) = e^x$, where e is the irrational number $2.718\ldots$, is called **the** exponential function.

At the point $(0, 1)$, where $y = e^x$ cuts the y-axis, the curve has gradient 1.

Note:
e^x is sometimes written exp (x).

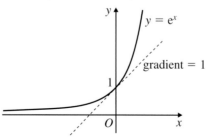

Note:
In fact, for all points on the curve $y = e^x$, $\dfrac{dy}{dx} = e^x$
(Section 3.1).

$e^x > 0$ for all values of x and, using the laws of indices, $e^0 = 1$.

Notice that as $x \to \infty$, $e^x \to \infty$ rapidly.

Recall:
$a^0 = 1$ (C1 Section 1.1).

Example 1.15 $f(x) = 2 + e^{-x}$.

a Sketch the graph of $y = f(x)$.

b State the equations of any asymptotes to the curve.

c State the domain and range of the function.

Tip:
The word 'state' means you don't need to calculate these answers or show any working; you should just be able to write them down.

Step 1: Decide the general shape of the curve, using your knowledge of transformations.

a Let $y = 2 + e^{-x}$

The graph of $y = e^{-x}$ is a reflection in the y-axis of $y = e^x$.

The graph of $y = 2 + e^{-x}$ is a translation of $y = e^{-x}$ by 2 units in the y-direction.

Step 2: Set $x = 0$ to find the y intercept.

When $x = 0$, $y = 2 + e^0 = 2 + 1 = 3$, so the curve cuts the y-axis at $(0, 3)$.

Recall:
$y = f(-x)$ is a reflection in the y-axis of $y = f(x)$, $y = f(x) + a$ translates the graph of $y = f(x)$ by a units in the y-direction (C1 Section 3.8).

Note:
e^{-x} is always greater than 0, so y has to be greater than $2 + 0$ i.e. $y > 2$.

Step 3: Sketch the curve, marking the intercept.

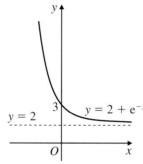

Tip:
As $x \to \infty$, $f(x) \to 2$, so don't let your graph turn upwards at the end.

Step 4: Write down the equation of the asymptote.

b The x-axis is an asymptote to $y = e^{-x}$, so $y = 2$ is an asymptote to the translated curve.

Step 5: Using the graph, write down the domain and range.

c The domain is $x \in \mathbb{R}$ and the range is $f(x) \in \mathbb{R}$, $f(x) > 2$.

Recall:
The domain is the set of values that x can take. The range is the set of values that y can take. Both can be determined from the graph (Section 1.1).

The logarithmic function ln x

In *Core 2* you used the fact that $x = a^y \Leftrightarrow \log_a x = y$ to show that logarithmic functions are the inverses of exponential functions. It follows that $x = e^y \Leftrightarrow \log_e x = y$.

The logarithm to the base e of x is called **ln x** and is the **natural** log of x. So

$$x = e^y \Leftrightarrow \ln x = y$$

Since $\ln x$ is the inverse of e^x, the graph of $y = \ln x$ is a reflection of the graph of $y = e^x$ in the line $y = x$.

The domain of $\ln x$ is the range of e^x, i.e. $x \in \mathbb{R}$, $x > 0$.

The range of $\ln x$ is the domain of e^x, i.e. $y \in \mathbb{R}$.

Recall:
Inverse functions (Section 1.3).

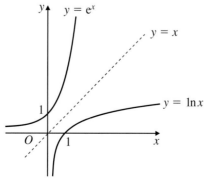

Using the laws of logs, $\ln 1 = 0$ and $\ln e = 1$.

Notice that as $x \to \infty$, $\ln x \to \infty$ slowly.

Recall:
$\log_a 1 = 0$, $\log_a a = 1$ (C2 Section 3.5).

Example 1.16 Describe the transformations required to map the graph of $y = \ln x$ onto the graph of $y = \frac{1}{2} \ln (x - 2) + 1$ and sketch the graph of $y = \frac{1}{2} \ln (x - 2) + 1$.

Step 1: Define the transformations.

The graph of $y = \frac{1}{2} \ln (x - 2) + 1$ is obtained from the graph of $y = \ln x$ by a stretch by scale factor $\frac{1}{2}$ parallel to the y-axis, followed by a translation by $\binom{2}{1}$.

Recall:
$y = a\mathrm{f}(x)$ is a stretch, scale factor a in the y-direction.
$y = \mathrm{f}(x - b) + c$ is a translation by $\binom{b}{c}$ (Section 1.4).

Step 2: Consider any asymptotes.

Since $y = \ln x$ has an asymptote at $x = 0$, $y = \frac{1}{2} \ln (x - 2) + 1$ must have an asymptote at $x = 2$.

Step 3: Draw in the asymptote.

Step 4: Sketch the curve taking into account the stretch and translation.

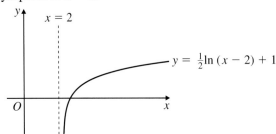

Note:
The stretch must be carried out first. If the order of the transformations is reversed, the resulting graph is $y = \frac{1}{2} \ln (x - 2) + \frac{1}{2}$.

Example 1.17 Given that $\mathrm{f}(x) = \ln (3 - x)$, $x \in \mathbb{R}$, $x < 3$, find an expression for $\mathrm{f}^{-1}(x)$, stating the domain and range of f^{-1}.

Tip:
Be careful!
$\ln (3 - x) \neq \ln 3 - \ln x$

Step 1: Let $y = \mathrm{f}(x)$ and interchange x and y.

Let $y = \ln (3 - x)$
Now let $x = \ln (3 - y)$

Recall:
$\ln y = x \Leftrightarrow y = e^x$

Step 2: Make y the subject.

$$e^x = 3 - y$$
$$y = 3 - e^x$$
So $\mathrm{f}^{-1}(x) = 3 - e^x$.

Note:
You can interchange x and y at the end if you prefer.

Step 3: State the domain and range of the inverse function.

Domain is $x \in \mathbb{R}$.
Range is $\mathrm{f}^{-1}(x) \in \mathbb{R}$, $\mathrm{f}^{-1}(x) < 3$.

Recall:
The domain of f^{-1} is the range of f. The range of f^{-1} is the domain of f (Section 1.3).

Solving equations involving e^x and $\ln x$

When solving equations involving e^x or $\ln x$, make use of the laws of indices and logarithms that you already know.

You will also need to use the fact that $\ln x$ is the inverse of e^x.

To solve an equation of the form $e^{ax + b} = p$, first take natural logs of both sides.

To solve an equation of the form $\ln (ax + b) = q$, first rewrite it as $ax + b = e^q$.

Also look out for cases when a simple substitution will transform the equation into a quadratic or cubic equation that is easy to solve.

Recall:
Laws of indices (C1 Section 1.1) and laws of logarithms (C2 Section 3.5).

Recall:
Natural logs are logs to the base e.

Example 1.18 Find the exact solutions of the following equations:

a $e^{6x - 1} = 3$

b $e^x = 6e^{-x} + 5$

Recall:
$\log a^n = n \log a$
(C2 Section 3.5).

Step 1: Take natural logs of both sides

Step 2: Use an appropriate log law.

a $e^{6x - 1} = 3$
$$\ln (e^{6x - 1}) = \ln 3$$
$$(6x - 1)\ln e = \ln 3$$
$$6x - 1 = \ln 3$$

Recall:
$\ln e = 1$

Step 3: Rearrange the equation to find x.

$$6x = \ln 3 + 1$$
$$x = \frac{\ln 3 + 1}{6}$$

Tip:
The word 'exact' here means leave your answers in terms of $\ln a$. Don't give your answer as a rounded decimal.

b $e^x = 6e^{-x} + 5$

Step 1: Substitute for e^x. Let $y = e^x$

Then $\qquad y = 6 \times \dfrac{1}{y} + 5 = \dfrac{6}{y} + 5$

Step 2: Multiply through by y.
$$y^2 = 6 + 5y$$

Step 3: Rearrange the equation and solve for y.
$$y^2 - 5y - 6 = 0$$
$$(y - 6)(y + 1) = 0$$
$$\Rightarrow \quad y = 6 \text{ or } y = -1$$

Step 4: Substitute back e^x for y. So $e^x = 6$ or $e^x = -1$.

Step 5: Solve for x, using $\ln x$ as the inverse of e^x. Since $e^x > 0$ for all values of x, $e^x = -1$ has no solution, but if $e^x = 6$, $x = \ln 6$.

Note:
If $y = e^x$, then $e^{-x} = \dfrac{1}{e^x} = \dfrac{1}{y}$.

Tip:
Don't forget to multiply **every** term by y.

Tip:
If the expression does not factorise use the quadratic formula.

Tip:
Don't forget to finish the question! Marks are often lost by candidates who forget to substitute back.

Example 1.19 Find the exact solutions of the following equations:

   a $\ln (2y + 1)^2 = 6$

   b $\ln (y + 1) - \ln y = 2$

Step 1: Use an appropriate log law.
  a $\quad \ln (2y + 1)^2 = 6$
$$2 \ln (2y + 1) = 6$$
$$\ln (2y + 1) = \dfrac{6}{2} = 3$$

Step 2: Change the log to exponential form.
$$2y + 1 = e^3$$

Step 3: Rearrange to find y.
$$2y = e^3 - 1$$
$$y = \dfrac{e^3 - 1}{2}$$

  b $\quad \ln (y + 1) - \ln y = 2$

Step 1: Simplify using the log laws.
$$\ln \dfrac{y + 1}{y} = 2$$

Step 2: Change the log to exponential form.
$$\dfrac{y + 1}{y} = e^2$$

Step 3: Eliminate the denominator.
$$y + 1 = e^2 y$$

Step 4: Make y the subject by rearranging and factorising.
$$e^2 y - y = 1$$
$$y (e^2 - 1) = 1$$
$$y = \dfrac{1}{e^2 - 1}$$

Recall:
$\log a^n = n \log a$

Tip:
$\ln y = x \Leftrightarrow y = e^x$

Note:
You don't have to simplify the log first; an alternative method would be
$(2y + 1)^2 = e^6$, so
$2y + 1 = \sqrt{e^6} = e^{\frac{6}{2}} = e^3$,
as before.

Recall:
$\log x - \log y = \log \dfrac{x}{y}$
(C2 Section 3.5).

Tip:
Be careful!
$\ln \dfrac{y + 1}{y} \neq \dfrac{\ln y + 1}{\ln y}$

Tip:
From your calculator
$\dfrac{1}{e^2 - 1} = 0.1565...$
Substitute back into the original equation to make sure you haven't made a slip.

In real life, exponential growth (e.g. population growth, investment growth, appreciation in value) is modelled around the exponential equation $y = Ae^{kt}$, where A and k are constants, $k > 0$.

Exponential decay (e.g. decay in radioactive isotopes, depreciation in value, temperature cooling) is modelled around the exponential equation $y = Ae^{-kt}$, where A and k are constants, $k > 0$.

Note:
This topic is extended to rates of increase and decrease in Section 3.2. It is further extended in C4.

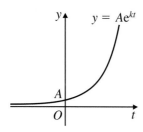
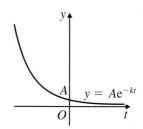

Example 1.20 The temperature, $T\,°C$, of a microwave meal t minutes after it has been heated is given by

$$T = 18 + 50e^{-\frac{t}{20}}, \; t \geq 0.$$

a Find, in °C, the temperature of the meal the instant that it has been heated.

b Calculate, in °C to three significant figures, the temperature of the meal 5 minutes after it has been heated.

c Calculate, to the nearest minute, the time at which the temperature of the meal is 45 °C.

Step 1: Substitute $t = 0$ to find T.

a When $t = 0$,

$$\begin{aligned} T &= 18 + 50e^{-\frac{0}{20}} \\ &= 18 + 50e^0 \\ &= 18 + 50 \times 1 \\ &= 68 \end{aligned}$$

So the temperature of the meal the instant that it has been heated is 68 °C.

Tip:
t is the time after the meal has been heated, so, when the meal comes out of the oven, $t = 0$.

Recall:
$e^0 = 1$

Step 2: Substitute the given value of t to find T.

b When $t = 5$,

$$\begin{aligned} T &= 18 + 50e^{-\frac{5}{20}} \\ &= 18 + 50e^{-\frac{1}{4}} \\ &= 18 + 50 \times 0.7788\ldots \\ &= 56.94\ldots \end{aligned}$$

So the temperature of the meal 5 minutes after it has been heated is 56.9 °C (3 s.f.).

Tip:
Use the full value in your calculation, but don't forget to give your answer to the required degree of accuracy.

Step 3: Substitute the given value of T into the equation.

Step 4: Simplify the equation.

c When $T = 45$,

$$45 = 18 + 50e^{-\frac{t}{20}}$$

$$27 = 50e^{-\frac{t}{20}}$$

$$\frac{27}{50} = e^{-\frac{t}{20}}$$

Note:
You need to isolate the exponential term ($e^{-\frac{t}{20}}$) before you take logs.

Step 5: Take natural logs of both sides and use an appropriate log law to simplify.

$$\ln \tfrac{27}{50} = \ln e^{-\frac{t}{20}}$$

$$= -\frac{t}{20} \ln e$$

$$= -\frac{t}{20}$$

Step 6: Rearrange to find *t* and evaluate using your calculator.

$$-20 \ln \tfrac{27}{50} = t$$

$$t = 12.3\ldots$$

Recall:
$\ln e = 1$

Tip:
Use the exact values in your working; only use decimals at the end.

So the temperature of the meal is 45 °C approximately 12 minutes after it is heated.

SKILLS CHECK **1C: Exponentials, logarithms and exponential growth and decay**

 1 Solve $e^{-\frac{x}{3}} = \tfrac{1}{2}$, giving your answer in the form $a \ln b$ where a and b are integers.

 2 Solve $3e^{2x} = 2(e^x + 4)$.

3 Solve $e^{2x-5} = 1$.

4 Find the exact solution of $\ln (4x + 3) = 0.5$.

5 Find the exact solutions of $(\ln x)^2 - 5 \ln x + 6 = 0$.

6 On separate sets of axes, sketch each of the following, stating the exact coordinates of any points of intersection with the *x*-axis and the equations of any asymptotes.

　a $y = |\ln x|, x > 0$

　b $y = \ln (x - 3) + 2, x > 3$

7 For each of the functions f defined by,

　a $f(x) = 3e^x, x \in \mathbb{R}$ 　　　　**b** $f(x) = \ln 2x, x \in \mathbb{R}, x > 0$

　i find an expression, in terms of *x*, for $f^{-1}(x)$,

　ii sketch the graphs of $y = f(x)$ and $y = f^{-1}(x)$ on the same axes, showing the coordinates of any points of intersection with the axes,

　iii state the range of f and the domain and range of f^{-1}.

 8 Given that $f(x) = \tfrac{1}{4} \ln (x + 1), x \in \mathbb{R}, x > -1$,

　a find an expression, in terms of *x*, for $f^{-1}(x)$,

　b state the domain and range of f^{-1},

　c find the exact value of *x* for which $f(x) = \tfrac{1}{2}$.

9 Given that $f(x) = e^{2x+1}, x \in \mathbb{R}$,

　a find an expression, in terms of *x*, for $f^{-1}(x)$,

　b sketch the graphs of $y = f(x)$ and $y = f^{-1}(x)$ on the same axes, stating the equations of any asymptotes,

　c state the range of f and the domain and range of f^{-1}.

10 The amount an initial investment of £1000 is worth after t years is given by A, where

$$A = 1000e^{0.09t}, \; t \geqslant 0.$$

a How much is the investment worth after 5 years?

b After how many years will the investment have doubled in value?

c Sketch the graph of A against t.

SKILLS CHECK **1C EXTRA** is on the CD

Examination practice 1 Algebra and functions

1 The function f is defined by

$$f: x \mapsto 1 - x^2, \quad x \leqslant 0.$$

i Sketch the graph of f.

ii Find an expression, in terms of x, for $f^{-1}(x)$ and state the domain of f^{-1}.

iii The function g is defined by

$$g: x \mapsto 2x, \quad x \leqslant 0.$$

Find the value of x for which $fg(x) = 0$. [OCR Nov 1997]

2 The function f is defined by

$$f: x \mapsto \frac{1}{\sqrt{x}} + 2, \quad x > 0.$$

i State the range of f. **ii** Find an expression for $f^{-1}(x)$. [OCR June 2003]

3 Express $x^2 + 4x$ in the form $(x + a)^2 + b$, stating the numerical values of a and b.
The functions f and g are defined as follows:

$$f: x \mapsto x^2 + 4x, \quad x \geqslant -2,$$
$$g: x \mapsto x + 6, \quad x \in \mathbb{R}.$$

i Show that the equation $gf(x) = 0$ has no real roots.

ii State the domain of f^{-1}.

iii Find an expression in terms of x for $f^{-1}(x)$.

iv Sketch, on a single diagram, the graphs of $y = f(x)$ and $y = f^{-1}(x)$. [OCR June 1996]

4 The function f is defined for all real values of x by

$$f(x) = 25 + \sqrt[3]{x}.$$

i Evaluate ff(8). **ii** Express $f^{-1}(x)$ in terms of x.

iii

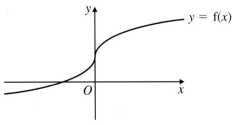

The graph of $y = f(x)$ is shown in the diagram. Copy the diagram and sketch the graph of $y = f^{-1}(x)$ on your copy. Indicate how the graphs of $y = f(x)$ and $y = f^{-1}(x)$ are related. [OCR Jan 2004]

5 The functions f and g are defined for all real values of x by

$$f(x) = 3x - 10,$$
$$g(x) = |x - 2|.$$

  i State the range of each function.

  ii Show that the value of $fg(-4)$ is 8.

  iii Determine the value of $f^{-1}(7)$.

  iv Solve the equation $gf(x) = 1$.   [OCR Nov 2003]

6 Find the exact solution of the equation

$$|7x - 3| = |7x + 6|.$$ [OCR June 2003]

7 **i** Solve the inequality $|x - 250| < 10$.

  ii Hence determine all the integers n which satisfy the inequality

$$|1.02^n - 250| < 10.$$ [OCR Jan 2004]

8 The functions f and g are defined as follows:

$$f: x \mapsto \frac{1}{x + 2}, \quad x > 0,$$

$$g: x \mapsto \frac{1}{x}, \quad\quad x > 0.$$

  i Find an expression in terms of x, for $f^{-1}(x)$ and state the domain of f^{-1}.

  ii By using the substitution $u = \sqrt{x}$, or otherwise, solve the equation $\dfrac{1}{f(x)} = 3\sqrt{x}$.

  iii Show that the equation $fg(x) = gf(x)$ has no solutions.   [OCR Nov 1996]

9 The function f is defined by

$$f: x \mapsto \ln(x + 1), \quad\quad x > -1.$$

Find an expression for $f^{-1}(x)$ and state the domain and range of the inverse function f^{-1}.

The function is defined by

$$g: x \mapsto x - 1, \quad\quad x \in \mathbb{R}.$$

Describe the geometrical relationship between the graphs of $y = fg(x)$ and $y = gf(x)$.

                                                   [OCR Nov 1998]

10

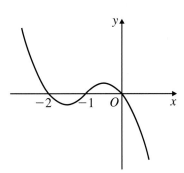

The graph of $y = f(x)$ is shown in the diagram. On separate diagrams sketch the graphs of

  i $y = f(x + 1)$,

  ii $y = |f(x + 1)|$,

showing the coordinates of the points where the graphs meet the x-axis.   [OCR June 1996]

11

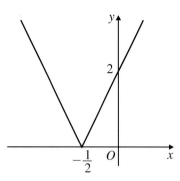

The diagram shows the graph of $y = |f(x)|$, where $f(x) = ax + b$. Find the possible values of the constants a and b, and hence give the two possible expressions for $f(x)$.

State the least value of x for which $|f(x)| \leqslant 2$. [OCR Nov 1999]

 12 The functions f and g are defined as follows, where a is a positive constant:

$$f: x \mapsto a - x, \qquad x \in \mathbb{R}$$
$$g: x \mapsto |2x + a|, \qquad x \in \mathbb{R}.$$

i Find $gf(4a)$.

ii On the same set of axes, sketch the graphs of $y = f(x)$ and $y = g(x)$.

iii Solve the equation $f(x) = g(x)$.

 13

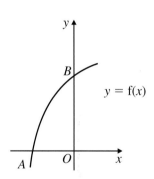

The diagram shows a sketch of the curve with equation $y = f(x)$. The curve crosses the x-axis at $A(-0.5, 0)$ and the y-axis at $B(0, 1)$.

Give the coordinates of A_1 and B_1, the images of A and B, when each of these transformations is applied:

a $y = 3f(2x)$ **b** $y = 2f(-x)$ **c** $y = f^{-1}(x)$

14 Describe the translation which maps the curve $y = \ln x$ onto the curve $y = \ln x + \ln 6$.

Express $\ln x + \ln 6$ as a single logarithm and hence describe an alternative transformation which maps the curve $y = \ln x$ onto the curve $y = \ln x + \ln 6$. [OCR Nov 1999]

15 i Sketch, on the same axes, the graphs of $y = \ln x$ and $y = \ln (x - a)$, where a is a positive constant. Label each graph, and indicate the coordinates of any points of intersection with the x-axis.

ii Given that $\ln x - \ln (x - a) = \ln 5$, find x in terms of a. [OCR Mar 2000]

16 It is given that $\ln x = p + 2$ and $\ln y = 3p$.

i Express each of the following in terms of p:

a $\ln (xy)$, **b** $\ln (x^3)$, **c** $\ln \left(\dfrac{y}{e} \right)$.

ii Express y in terms of x and e, simplifying your answer. [OCR Nov 2003]

17 Solve the following equations, giving the exact values of x and y:

a $e^{-2x} = \frac{1}{16}$

b $\ln y - \ln (y - 1) = 1$

18 Find the exact solutions of the following equations:

a $e^{3x + 6} = 8$

b $\ln (2x - 4) = 3$

 19 i Sketch, on the same axes, the graphs with equations

$$y = 1 + e^{-x} \quad \text{and} \quad y = 2|x + 4|$$

ii Write down the coordinates of any points where the graphs meet the axes.

The graphs intersect at the point where $x = p$.

iii Show that $x = p$ is a root of the equation $e^{-x} - 2x - 7 = 0$.

20 A curve has equation $y = f(x)$ where the function f is given by

$$f(x) = e^{x + 1} - 2, \qquad x \in \mathbb{R}$$

i Sketch the curve and write down the exact coordinates of the points of intersection with the axes.

ii Find the inverse function, f^{-1}.

iii State the domain and range of f^{-1}.

21 i Sketch the graph of $y = k + \ln \frac{1}{2} x$.

ii Find, in terms of k, the coordinates of the point of intersection with the x-axis.

iii Given that the curve crosses the x-axis at the point $\left(\dfrac{2}{e^2}, 0\right)$, show that $k = 2$.

22 The functions f and g are defined as follows:

$$\text{f}: x \mapsto 5e^{\frac{1}{2}x} - 6, \quad x \in \mathbb{R},$$
$$\text{g}: x \mapsto |x - 2|, \quad x \in \mathbb{R}.$$

i Determine the range of f and the range of g.

ii Find an expression for $f^{-1}(x)$.

iii Find the solutions of the equation $gf(x) = 7$, giving each in an exact form. [OCR Jan 2002]

2 Trigonometry

2.1 Inverse trigonometric functions

Use the notation $\sin^{-1}x$, $\cos^{-1}x$, $\tan^{-1}x$ to denote the principal values of the inverse trigonometric relations and relate their graphs (for the appropriate domains) to those of sine, cosine and tangent.

The three trigonometric functions of sine, cosine and tangent are many-one functions, since, for example,

$$\sin 30° = \sin 150° = \sin 390° = ... = 0.5.$$

Since they are many-one functions, they do not have an inverse.

However, if we restrict the domain, it is possible to define the **inverse trigonometric functions**: $\sin^{-1}x$, $\cos^{-1}x$ and $\tan^{-1}x$, also written $\arcsin x$, $\arccos x$ and $\arctan x$.

The graphs of the inverse functions are shown below. They have been drawn using the following properties of functions:

For a function f and its inverse function f^{-1}:

- the domain of f^{-1} is the range of f
- the range of f^{-1} is the domain of f
- the graph of $y = f^{-1}(x)$ is a reflection in the line $y = x$ of the graph of $y = f(x)$.

Recall:
Only one-one functions have an inverse (Section 1.3).

Note:
You can use degrees or radians.

Note:
To show the reflection, the scales on both axes must be the same, so radians are used in the diagrams below.

$\sin^{-1}x$

| f: $x \mapsto \sin x$ | Reflect $y = \sin x$ in the line $y = x$ | $f^{-1}: x \mapsto \sin^{-1}x$ |

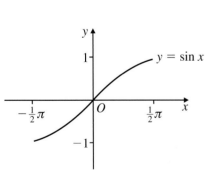
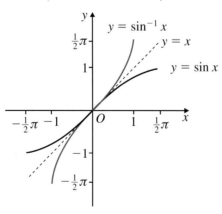
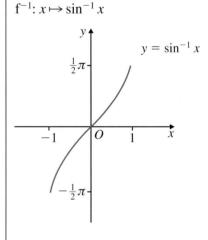

Domain $\quad -\frac{1}{2}\pi \leqslant x \leqslant \frac{1}{2}\pi$
Range $\quad -1 \leqslant \sin x \leqslant 1$

Domain $\quad -1 \leqslant x \leqslant 1$
Range $\quad -\frac{1}{2}\pi \leqslant \sin^{-1}x \leqslant \frac{1}{2}\pi$

Notice that if you turn your page through a quarter turn clockwise, and imagine the axis that is now horizontal as the x-axis, you can see a reflection in that axis of the sine curve.

$\cos^{-1} x$

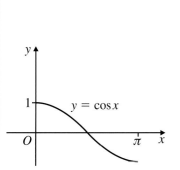

f: $x \mapsto \cos x$

Domain $0 \leqslant x \leqslant \pi$
Range $-1 \leqslant \cos x \leqslant 1$

Reflect $y = \cos x$ in the line $y = x$.

$y = \cos^{-1} x$
$y = x$
$y = \cos x$

f^{-1}: $x \mapsto \cos^{-1} x$

$y = \cos^{-1} x$

Domain $-1 \leqslant x \leqslant 1$
Range $0 \leqslant \cos^{-1} x \leqslant \pi$

Notice that if you turn your page through a quarter turn clockwise, and imagine the axis that is now horizontal as the x-axis, you can see a reflection in that axis of the cosine curve.

$\tan^{-1} x$

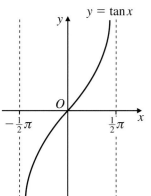

f: $x \mapsto \tan x$

$y = \tan x$

Domain $-\frac{1}{2}\pi < x < \frac{1}{2}\pi$
Range $\tan x \in \mathbb{R}$

Reflect $y = \tan x$ in the line $y = x$.

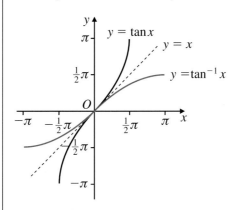

$y = \tan x$
$y = x$
$y = \tan^{-1} x$

f^{-1}: $x \mapsto \tan^{-1} x$

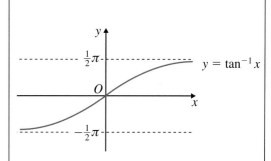

$y = \tan^{-1} x$

Domain $x \in \mathbb{R}$
Range $-\frac{1}{2}\pi < \tan^{-1} x < \frac{1}{2}\pi$

The lines $y = -\frac{1}{2}\pi$ and $y = \frac{1}{2}\pi$ are asymptotes to the curve $y = \tan^{-1} x$.

Notice that if you turn your page through a quarter turn clockwise, and imagine the axis that is now horizontal as the x-axis, you can see a reflection in that axis of the tan curve.

Principal value (PV)

Remember that the value of $\sin^{-1} x$, $\cos^{-1} x$ and $\tan^{-1} x$ is an **angle**. It is the value given on the calculator by the inverse trig functions and is often referred to as the **principal value** (PV).

Recall:
The PV is used when solving trig equations. You can use degrees or radians (C2 Section 1.7).

Example 2.1   **a**   Use a calculator to find, in degrees, the value of:

   i $\sin^{-1} 1$ **ii** $\cos^{-1} 0.5$

   iii $\tan^{-1}(-1)$ **iv** $\cos^{-1}(-0.3)$

b   Use a calculator to find, in radians correct to two decimal places, the value of:

   i $\sin^{-1}(-0.6)$

   ii $\cos^{-1} 0.3$

   iii $\tan^{-1} 10$

Step 1: Use appropriate inverse trig functions on the calculator.

a   **i** $\sin^{-1} 1 = 90°$ **ii** $\cos^{-1} 0.5 = 60°$

   iii $\tan^{-1}(-1) = -45°$ **iv** $\cos^{-1}(-0.3) = 107.45...°$

b   **i** $\sin^{-1}(-0.6) = -0.6435... = -0.64^c$ (2 d.p.)

   ii $\cos^{-1} 0.3 = 1.2661... = 1.27^c$ (2 d.p.)

   iii $\tan^{-1} 10 = 1.4711... = 1.47^c$ (2 d.p.)

> **Tip:**
> For part **a**, set your calculator to degree mode.

> **Tip:**
> For part **b**, set your calculator to radians mode.

Special angles

It is useful to learn the values of sin, cos and tan of these special angles. They are especially useful if you are asked to give *exact* values, in terms of π, for $\sin^{-1} x$, $\cos^{-1} x$ and $\tan^{-1} x$.

> **Recall:**
> Special angles (C2 Section 1.5).

For example,

$$\sin^{-1} \frac{1}{\sqrt{2}} = \frac{1}{4}\pi$$

$$\cos^{-1}\left(-\frac{1}{2}\right) = \frac{2}{3}\pi$$

$$\tan^{-1} \frac{1}{\sqrt{3}} = \frac{1}{6}\pi$$

$x°$	x^c	$\sin x$	$\cos x$	$\tan x$
30°	$\frac{1}{6}\pi$	$\frac{1}{2}$	$\frac{\sqrt{3}}{2}$	$\frac{1}{\sqrt{3}}$
45°	$\frac{1}{4}\pi$	$\frac{1}{\sqrt{2}}$	$\frac{1}{\sqrt{2}}$	1
60°	$\frac{1}{3}\pi$	$\frac{\sqrt{3}}{2}$	$\frac{1}{2}$	$\sqrt{3}$

Calculator note

When finding $\sin^{-1} x$, $\cos^{-1} x$ or $\tan^{-1} x$, the value given by the calculator in radians is a decimal. To write it as a multiple of π, divide the decimal value by π, then check whether the multiple can be written as a fraction by pressing the fraction key. You must beware, however! This method will probably not work if you have already rounded values prior to this in a calculation. It is better to learn and be able to recognise the trigonometric ratios of the special angles.

> **Tip:**
> This is useful if you forget the values of sin, cos and tan of the special angles, but do not rely on it as a method.

2.2 Reciprocal trigonometric functions

Understand the relationship of the secant, cosecant and cotangent functions to cosine, sine and tangent, and use properties and graphs of all six trigonometric functions for angles of any magnitude.

The **reciprocal functions** of the three main trig functions of cosine (cos), sine (sin) and tangent (tan) are **secant** (sec), **cosecant** (cosec) and **cotangent** (cot). They are defined as follows:

$$\sec x = \frac{1}{\cos x} \qquad \operatorname{cosec} x = \frac{1}{\sin x} \qquad \cot x = \frac{1}{\tan x}$$

Since $\tan x = \dfrac{\sin x}{\cos x}$, we can also write $\cot x = \dfrac{\cos x}{\sin x}$.

> **Note:**
> Do not confuse the notation for the *reciprocal* function with the *inverse* function. For example, cosec x can be written $(\sin x)^{-1}$ whereas the inverse sine function is written $\sin^{-1} x$.

Graphs of reciprocal functions

You should learn the properties of the graphs of $y = \sec x$, $y = \operatorname{cosec} x$ and $y = \cot x$.

Note:

A graphical calculator may be used in C3.

It is useful to understand how they are obtained from the graphs of $y = \sin x$, $y = \cos x$ and $y = \tan x$ and these are also shown in the diagrams below.

$y = \sec x$

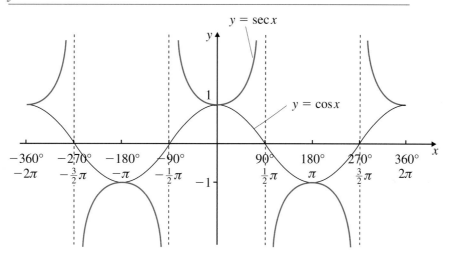

The value of $\sec x$ is always greater than or equal to 1 or less than or equal to -1.

There are vertical asymptotes through the points where $y = \cos x$ crosses the x-axis.

There are minimum points where $y = \cos x$ has maximum points.

There are maximum points where $y = \cos x$ has minimum points.

$y = \operatorname{cosec} x$

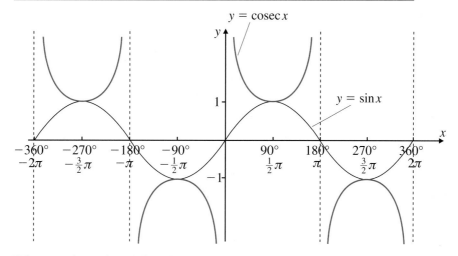

Whereas $\sin x$ takes values between -1 and 1, $\operatorname{cosec} x$ is always greater than or equal to 1 or less than or equal to -1.

There are vertical asymptotes through the points where $y = \sin x$ crosses the x-axis.

There are minimum points where $y = \sin x$ has maximum points.

There are maximum points where $y = \sin x$ has minimum points.

$y = \cot x$

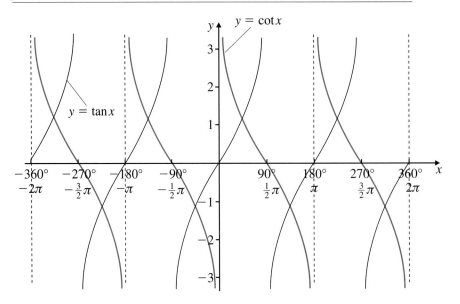

$\cot x$ can take all values.

The graph of $y = \cot x$ has vertical asymptotes through the points where $y = \tan x$ crosses the x-axis.

The graph of $y = \cot x$ crosses the x-axis where $y = \tan x$ has vertical asymptotes.

Applying transformations

You could be asked to apply combinations of transformations to the graphs of $y = \sec x$, $y = \operatorname{cosec} x$ and $y = \cot x$.

Recall:
Transformations (Section 1.4).

Example 2.2 $f(x) = 2\sec x + 1$.

    a The graph of $y = f(x)$ can be obtained from the graph of $y = \sec x$ by applying a stretch followed by a translation.

       i State the scale factor and direction of the stretch.

      ii Describe the translation.

     iii State the coordinates of the image of the point $(0, 1)$ under this mapping.

    b Write down the range of f for $-90° < x < 90°$.

Step 1: Compare with $y = af(x) + b$ and describe the stretch and the translation

Step 2: Apply the transformations in turn.

Step 3: Identify the range of f from the information in **a**

a i The stretch is in the y-direction and the scale factor is 2.

  ii The translation is by 1 unit in the y-direction.

 iii Under the stretch, the image of $(0, 1)$ is $(0, 2)$.

     Under the translation, the image of $(0, 2)$ is $(0, 3)$.

     So the image of $(0, 1)$ is $(0, 3)$.

b The range of f for $-90° < x < 90°$ is $f(x) \geqslant 3$.

Note:
The vector of the translation is $\binom{0}{1}$.

Tip:
For $-90° < x < 90°$, the minimum point of $y = \sec x$ is $(0, 1)$ and the range is $\sec x \geqslant 1$. What is the minimum point of $y = 2\sec x + 1$?

Use trigonometrical identities for the simplification and exact evaluation of expressions, and in the course of solving equations within a specified interval, and to select an identity or identities appropriate to the context. Show familiarity with the use of $\sec^2 \theta = 1 + \tan^2 \theta$ and $\operatorname{cosec}^2 \theta = 1 + \cot^2 \theta$

In *Core 2* you used this identity:

$$\cos^2 \theta + \sin^2 \theta \equiv 1 \qquad \text{①}$$

Manipulating this identity gives two further identities which must be learnt for *Core 3*:

$$1 + \tan^2 \theta = \sec^2 \theta \qquad \text{②}$$

$$\cot^2 \theta + 1 \equiv \operatorname{cosec}^2 \theta \qquad \text{③}$$

Remember that these are identities, so they are true for all values of θ, in degrees or in radians.

You may be asked to use these identities to prove further identities or solve equations as in the following examples.

Recall:
Trig identities (C2 Section 1.6).

Tip:
For ②, divide each term of identity ① by $\cos^2 \theta$.

Tip:
For ③, divide each term of identity ① by $\sin^2 \theta$.

Example 2.3 Solve the equation

$$\sec^2 \theta = 5(\tan \theta - 1),$$

where $0° \leqslant \theta \leqslant 360°$, giving your answers in degrees, correct to the nearest degree.

Step 1: Use an appropriate identity to form an equation in $\tan \theta$.

Tip:
Set your calculator to degrees mode.

$$\sec^2 \theta = 5(\tan \theta - 1)$$
$$\Rightarrow \quad 1 + \tan^2 \theta = 5 \tan \theta - 5$$
$$\Rightarrow \quad \tan^2 \theta - 5 \tan \theta + 6 = 0$$

Step 2: Solve the equation in $\tan \theta$.

$$(\tan \theta - 3)(\tan \theta - 2) = 0$$
$$\Rightarrow \quad \tan \theta - 3 = 0$$
$$\tan \theta = 3$$
$$\theta = 71.5...°,$$
$$251.5...°$$

or
$$\tan \theta - 2 = 0$$
$$\tan \theta = 2$$
$$\theta = 63.4...°,$$
$$243.4...°$$

Tip:
The linear term is $\tan \theta$, so write $\sec^2 \theta$ in terms of $\tan \theta$.

Tip:
This is a quadratic equation in $\tan \theta$.

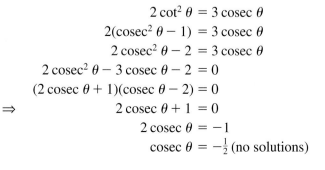

Tip:
Since the tan function repeats every 180°, the values of θ in range are PV and PV + 180° (C2 Section 1.7).

So, to the nearest degree $\theta = 63°, 72°, 243°, 252°$.

Example 2.4 Find the exact values of θ, where $-\pi \leqslant \theta \leqslant \pi$, such that

$$2 \cot^2 \theta = 3 \operatorname{cosec} \theta$$

Tip:
Notice the description *exact*.

Step 1: Use an appropriate identity to form an equation in $\operatorname{cosec} \theta$.

$$2 \cot^2 \theta = 3 \operatorname{cosec} \theta$$
$$2(\operatorname{cosec}^2 \theta - 1) = 3 \operatorname{cosec} \theta$$
$$2 \operatorname{cosec}^2 \theta - 2 = 3 \operatorname{cosec} \theta$$

Step 2: Factorise and solve.

$$2 \operatorname{cosec}^2 \theta - 3 \operatorname{cosec} \theta - 2 = 0$$
$$(2 \operatorname{cosec} \theta + 1)(\operatorname{cosec} \theta - 2) = 0$$
$$\Rightarrow \quad 2 \operatorname{cosec} \theta + 1 = 0$$
$$2 \operatorname{cosec} \theta = -1$$
$$\operatorname{cosec} \theta = -\tfrac{1}{2} \text{ (no solutions)}$$

Tip:
The linear term is $\operatorname{cosec} \theta$, so aim to form an equation just in $\operatorname{cosec} \theta$ using the relationship $1 + \cot^2 \theta \equiv \operatorname{cosec}^2 \theta$.

or $\cosec \theta - 2 = 0$

$\cosec \theta = 2$

$\sin \theta = \frac{1}{2}$

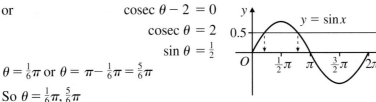

$\theta = \frac{1}{6}\pi$ or $\theta = \pi - \frac{1}{6}\pi = \frac{5}{6}\pi$

So $\theta = \frac{1}{6}\pi, \frac{5}{6}\pi$

Example 2.5 Prove the identity

$$(\sec x - \cosec x)(\sec x + \cosec x) \equiv (\tan x - \cot x)(\tan x + \cot x).$$

General steps for proving an identity:
Step 1: Start with one side of the identity and simplify/use appropriate identities to write it in a different format.
Step 2: Continue rewriting/simplifying until you get the expression on the other side of the identity.

$\text{LHS} = (\sec x - \cosec x)(\sec x + \cosec x)$

$= \sec^2 x - \cosec^2 x$

$= (1 + \tan^2 x) - (1 + \cot^2 x)$

$= 1 + \tan^2 x - 1 - \cot^2 x$

$= \tan^2 x - \cot^2 x$

$= (\tan x - \cot x)(\tan x + \cot x)$

$= \text{RHS}$

Note:
In a proof, all working must be shown.

Recall:
Factorising the difference of two squares, where
$(a - b)(a + b) = a^2 - b^2$
(C1 Section 2.4).

So $(\sec x - \cosec x)(\sec x + \cosec x) \equiv (\tan x - \cot x)(\tan x + \cot x)$.

When proving an identity, you can start with either side. So you can start with the left-hand side and aim to get to the expression on the right-hand side, or vice versa. This gives a neat method of proof, but it is also acceptable to show that each side is equal to the same (third) expression. This is sometimes referred to as 'meeting in the middle'.

SKILLS CHECK **2A: Trigonometric functions and identities**

1 Use your calculator to find the value, to the nearest degree, of each of the following:

  a $\cos^{-1} 0.45$ **b** $\sin^{-1}(-0.67)$ **c** $\tan^{-1} 2.8$ **d** $\cos^{-1}(-0.2)$

2 Find the exact value, in radians in terms of π, of each of the following:

  a $\tan^{-1} 1$ **b** $\sin^{-1}(-0.5)$ **c** $\cos^{-1} 0$ **d** $\cos^{-1}\left(\dfrac{1}{\sqrt{2}}\right)$

3 Solve the following equations, where $0° < x < 360°$. If your answer is not exact, give it to the nearest $0.1°$.

  a $\cosec x = 2.5$ **b** $\cot x = -\sqrt{3}$ **c** $\sec 2x = 1.5$ **d** $\sec^2 x = 4$

4 Solve the following equations, where $0 \leq \theta \leq 2\pi$. If your answer is not exact, give it to three significant figures.

  a $\cot \frac{1}{2}\theta = 0.4$ **b** $\cosec \theta = -1$ **c** $\sec(\theta + \frac{1}{2}\pi) = 4$ **d** $\cosec^2 \theta = 2$

 5 Prove these identities:

  a $\tan \theta + \cot \theta \equiv \sec \theta \cosec \theta$ **b** $\sec \theta - \cos \theta \equiv \tan \theta \sin \theta$

 6 Solve $\tan^2 \theta = 1 + \sec \theta$ for $0° \leq \theta \leq 360°$.

7 Solve $4 \cosec \theta - 5 = \cot^2 \theta$, for $-360° \leq \theta \leq 360°$.

8 Solve $\cot \theta = 2 \cos \theta$ for $0 < \theta \leq 2\pi$, giving your answers exactly in terms of π.

9 Find the exact values of θ, where $-\pi \leqslant \theta \leqslant \pi$, such that $\mathrm{cosec}^2\,\theta = 2\cot\theta$.

10 $\mathrm{f}(x) = -\cot(x + 90°)$.

    a Describe how the graph of $y = \mathrm{f}(x)$ may be obtained from the graph of $y = \cot x$.

    b On the same set of axes, sketch the graph of $y = \cot x$ and $y = -\cot(x + 90°)$ for $-180° \leqslant x \leqslant 180°$, labelling each curve clearly.

    c Describe the relationship between $y = -\cot(x + 90°)$ and $y = \tan x$.

11 $\mathrm{f}(x) = 2\,\mathrm{cosec}\,x + 1$.

    a Describe transformations that map the graph of $y = \mathrm{cosec}\,x$ onto the graph of $y = \mathrm{f}(x)$.

    b State the coordinates of the image of the point $\left(\frac{1}{2}\pi, 1\right)$ under this mapping.

    c Write down the range of f for $0 < x < \pi$.

12 **a** Describe transformations that map the graph of $y = \sec x°$ onto the graph of $y = \sec 2x° + 1$.

    b State the image of $(360, 1)$ under the mapping.

SKILLS CHECK **2A EXTRA** is on the CD

2.4 Addition and double angle formulae

Show familiarity with the expansions of $\sin(A \pm B)$, $\cos(A \pm B)$ and $\tan(A \pm B)$, and the formulae for $\sin 2A$, $\cos 2A$ and $\tan 2A$.

The addition formulae

The addition formulae give expressions for sin, cos and tan of the **sum** or **difference** of two angles A and B. They are identities and so are true for all values of A and B.

The six formulae are summarised in your formulae booklet as follows:

$$\sin(A \pm B) = \sin A \cos B \pm \cos A \sin B$$
$$\cos(A \pm B) = \cos A \cos B \mp \sin A \sin B$$
$$\tan(A \pm B) = \frac{\tan A \pm \tan B}{1 \mp \tan A \tan B}$$

> **Note:**
> Despite their name, the addition formulae relate to two angles added or subtracted, not just added. They are also called compound angle formulae.

> **Tip:**
> Make sure that you understand how to use the $\pm$ and $\mp$ symbols, for example
> $\cos(A - B)$
> $\equiv \cos A \cos B + \sin A \sin B$.

Example 2.6 **a** Given that $\cos(\theta + 30°) = \sin\theta$, show that $\tan\theta = \dfrac{1}{\sqrt{3}}$.

    b Hence find all the values of θ, where $0° < \theta < 360°$, for which

        $\cos(\theta + 30°) = \sin\theta$.

a $\cos(\theta + 30°) = \sin\theta$

Step 1: Expand the sum using the addition formula.
$$\Rightarrow \quad \cos\theta\cos 30° - \sin\theta\sin 30° = \sin\theta$$

Step 2: Insert known trig values.
$$\Rightarrow \quad \cos\theta \times \frac{\sqrt{3}}{2} - \sin\theta \times \frac{1}{2} = \sin\theta$$

> **Recall:**
> Trigonometric ratios of special angles (Section 2.1).

Step 3: Simplify to the required form.

Multiplying all the terms by 2 gives

$$\sqrt{3}\cos\theta - \sin\theta = 2\sin\theta$$
$$\sqrt{3}\cos\theta = 3\sin\theta$$

Dividing each term by $3\cos\theta$ gives

$$\frac{\sqrt{3}\cos\theta}{3\cos\theta} = \frac{3\sin\theta}{3\cos\theta}$$

$$\frac{\sqrt{3}}{3} = \tan\theta$$

$$\tan\theta = \frac{\sqrt{3}}{3} \times \frac{\sqrt{3}}{\sqrt{3}} = \frac{1}{\sqrt{3}}, \text{ as required}$$

Step 4: Use the result in **a** to solve the simple trig equation

b $\cos(\theta + 30°) = \sin\theta$

$$\Rightarrow \quad \tan\theta = \frac{1}{\sqrt{3}}$$

$$\theta = 30°, 210°$$

Note:
It is permissible to divide by $\cos\theta$ here, since $\cos\theta = 0$ is clearly not a solution of the equation.

Recall:
$\dfrac{\sin\theta}{\cos\theta} \equiv \tan\theta$

Tip:
Manipulate the surd to get the required format.

Tip:
The values in range are PV and PV + 180°. Remember that the tan function repeats every 180°.

Example 2.7 Prove the identity

$$\cos A \cos(A - B) + \sin A \sin(A - B) \equiv \cos B$$

Step 1: Expand the LHS using the addition formulae.

$$\begin{aligned}
\text{LHS} &= \cos A(\cos A \cos B + \sin A \sin B) \\
&\quad + \sin A(\sin A \cos B - \cos A \sin B) \\
&= \cos^2 A \cos B + \cos A \sin A \sin B + \sin^2 A \cos B \\
&\quad - \sin A \cos A \sin B
\end{aligned}$$

Step 2: Simplify and use appropriate identities to arrive at the RHS.

$$\begin{aligned}
&= \cos^2 A \cos B + \sin^2 A \cos B \\
&= \cos B(\cos^2 A + \sin^2 A) \\
&= \cos B \\
&= \text{RHS}
\end{aligned}$$

So $\cos A \cos(A - B) + \sin A \sin(A - B) \equiv \cos B$.

Tip:
You will not gain any marks if all you do is quote the formula for $\cos(A - B)$ and $\sin(A - B)$ given in the formulae booklet. You must then go on to use them in the question.

Tip:
$\cos^2 A + \sin^2 A \equiv 1$

Tip:
Keep going, even when the expression looks very complicated. Often terms cancel or you can use other identities to simplify further.

The double angle formulae

Putting $B = A$ in the identities for the sum of two angles gives the **double angle formulae** which are true for all values of A.

$$\begin{aligned}
\sin 2A &\equiv 2\sin A \cos A \\
\cos 2A &\equiv \cos^2 A - \sin^2 A \\
\tan 2A &\equiv \frac{2\tan A}{1 - \tan^2 A}
\end{aligned}$$

Often alternative formats of the formula for $\cos 2A$ are used. Using $\sin^2 A + \cos^2 A \equiv 1$

$$\begin{aligned}
\cos 2A &\equiv \cos^2 A - \sin^2 A \\
&\equiv \cos^2 A - (1 - \cos^2 A) \\
&\equiv 2\cos^2 A - 1
\end{aligned}$$

Also $\quad \begin{aligned}[t]
\cos 2A &\equiv \cos^2 A - \sin^2 A \\
&\equiv (1 - \sin^2 A) - \sin^2 A \\
&\equiv 1 - 2\sin^2 A
\end{aligned}$

Summarising:

$$\cos 2A \equiv \cos^2 A - \sin^2 A \equiv 2\cos^2 A - 1 \equiv 1 - 2\sin^2 A$$

Note:
You should learn these. However, if you forget them, put $B = A$ in the formulae for $\sin(A + B)$, $\cos(A + B)$ and $\tan(A + B)$, given in the formulae booklet.

Note:
Do not expect always to see $2A$ and A as the angles. Look out especially for **half angles**, where, for example

$\sin A \equiv 2\sin\dfrac{A}{2}\cos\dfrac{A}{2}$

$\cos A \equiv 2\cos^2\dfrac{A}{2} - 1$

Note:
You must be confident about using any of these formats.

Example 2.8 Solve the equation $\cos 2x + 3 \sin x = 2$ for $0° \leqslant x < 360°$.

Step 1: Use an appropriate double angle formula.

$$\cos 2x + 3 \sin x = 2$$
$$\Rightarrow \quad 1 - 2 \sin^2 x + 3 \sin x = 2$$

Step 2: Solve the equation in $\sin x$.

$$2 \sin^2 x - 3 \sin x + 1 = 0$$
$$(2 \sin x - 1)(\sin x - 1) = 0$$
$$\Rightarrow \quad 2 \sin x - 1 = 0 \qquad \text{or} \quad \sin x - 1 = 0$$
$$\sin x = 0.5 \qquad \qquad \sin x = 1$$
$$x = 30°, 150° \qquad \qquad x = 90°$$

So $x = 30°, 90°, 150°$

Tip:
Form an equation in $\sin x$ by using the version of $\cos 2x$ that involves $\sin x$.

Tip:
PV $= 30°$; another solution in range is $180° -$ PV.

Tip:
PV $= 90°$. This is the only value in range. Be careful not to include other incorrect values within the range which will lead to loss of marks.

Example 2.9 Given that $\tan x = \frac{3}{4}$, using an appropriate double angle formula find the *exact* value of $\cot 2x$.

Step 1: Expand using an appropriate double angle formula.

First find the value of $\tan 2x$.

$$\tan 2x \equiv \frac{2 \tan x}{1 - \tan^2 x}$$

Step 2: Substitute the known value.

$$= \frac{2 \times \frac{3}{4}}{1 - \left(\frac{3}{4}\right)^2} = \frac{24}{7}$$

Step 3: Use the reciprocal function.

$$\cot 2x = \frac{1}{\tan 2x} = 1 \div \frac{24}{7} = \frac{7}{24}$$

Tip:
You will gain no marks for calculating $\tan^{-1}\left(\frac{3}{4}\right)$ and using the angle obtained to calculate $\cot 2x$.

Tip:
If you use the fraction key on the calculator, you may need to put $\frac{3}{4}$ in a bracket before squaring.

Example 2.10 For values of x in the interval $0 \leqslant x \leqslant 2\pi$, solve the following equations, giving your answers exactly in terms of π.

a $\sin x = \sin 2x$ $\qquad$ **b** $\sin \left(\frac{1}{2}x\right) = \sin x$

Step 1: Use an appropriate double angle formula

a
$$\sin x = \sin 2x$$
$$\Rightarrow \quad \sin x = 2 \sin x \cos x$$

Step 2: Factorise and solve.

$$2 \sin x \cos x - \sin x = 0$$
$$\sin x(2 \cos x - 1) = 0$$
$$\Rightarrow \quad \sin x = 0 \qquad \text{or} \quad 2 \cos x - 1 = 0$$
$$x = 0, \pi, 2\pi \qquad \qquad \cos x = 0.5$$
$$x = \tfrac{1}{3}\pi, \tfrac{5}{3}\pi$$

So $x = 0, \frac{1}{3}\pi, \pi, \frac{5}{3}\pi, 2\pi$.

Tip:
Do not divide through by $\sin x$ as this will result in the loss of some solutions. Take it out as a factor.

Step 3: Notice the link with part **a** and apply a suitable substitution.

b Letting $\frac{1}{2}x = \theta$, the equation
$$\Rightarrow \quad \sin \left(\tfrac{1}{2}x\right) = \sin x, \qquad 0 \leqslant x \leqslant 2\pi$$
becomes $\quad \sin \theta = \sin 2\theta \qquad 0 \leqslant \theta \leqslant \pi$

Step 4: Solve, using your answers from **a**.

From **a**, $\qquad \theta = 0, \tfrac{1}{3}\pi, \pi$
$$\Rightarrow \quad \tfrac{1}{2}x = 0, \tfrac{1}{3}\pi, \pi$$
$$x = 0, \tfrac{2}{3}\pi, 2\pi$$

Tip:
Be on the look out for links between parts of questions.

Tip:
Double your answers from part **a**, but only include values in the appropriate range.

Example 2.11 **a** By expanding $\sin (2A + A)$ prove that
$$\sin 3A \equiv 3 \sin A - 4 \sin^3 A.$$

b Hence, or otherwise, solve
$$3 \sin x - 4 \sin^3 x = 1$$
for values of x such that $0° < x < 180°$.

Step 1: Use the addition formula suggested.

a $\sin 3A \equiv \sin (2A + A) \equiv \sin 2A \cos A + \cos 2A \sin A$

$ \equiv 2 \sin A \cos A \cos A + (1 - 2 \sin^2 A)\sin A$

$ \equiv 2 \sin A \cos^2 A + \sin A - 2 \sin^3 A$

Step 2: Use appropriate formulae to express all terms in terms of $\sin A$.

$ \equiv 2 \sin A(1 - \sin^2 A) + \sin A - 2 \sin^3 A$

$ \equiv 2 \sin A - 2 \sin^3 A + \sin A - 2 \sin^3 A$

$ \equiv 3 \sin A - 4 \sin^3 A$

Recall:
Formulae for $\sin(A + B)$, $\sin 2A$ and $\cos 2A$, $\cos^2 A$.

Step 3: Use the result in **a** to form a simple equation and solve in the appropriate range.

b $3 \sin x - 4 \sin^3 x = 1$

$\Rightarrow \qquad \sin 3x = 1$

$\Rightarrow \qquad\quad 3x = 90°, 450°$

$\qquad\qquad\quad x = 30°, 150°$

Range for $3x$:

$0° < x < 180°$

$0° < 3x < 540°$

Tip:
Make sure that you have included all the solutions in range.

Example 2.12 Prove that $\dfrac{\sin 2A}{1 - \cos 2A} \equiv \cot A$.

$$\text{LHS} = \frac{\sin 2A}{1 - \cos 2A}$$

Step 1: Use an appropriate double angle formula in the numerator.

$$= \frac{2 \sin A \cos A}{1 - \cos 2A}$$

$$= \frac{2 \sin A \cos A}{1 - (1 - 2 \sin^2 A)}$$

Step 2: Use an appropriate double angle formula in the denominator.

$$= \frac{2 \sin A \cos A}{2 \sin^2 A}$$

$$= \frac{\cos A}{\sin A}$$

$$= \cot A$$

$$= \text{RHS}$$

So $\dfrac{\sin 2A}{1 - \cos 2A} \equiv \cot A$

Tip:
In the numerator, use $\sin 2A = 2 \sin A \cos A$.

Tip:
In the denominator, use $\cos 2A = 1 - 2 \sin^2 A$.

Tip:
Do this in stages, one thing at a time. Do not try to do too many things at once.

SKILLS CHECK **2B: Addition and double angle formulae**

1 Given that angles A and B are acute and that $\sin A = \frac{4}{5}$ and $\cos B = \frac{12}{13}$, find the exact values of:

   a $\cos A$ **b** $\sin B$ **c** $\cos (A - B)$ **d** $\sec (A - B)$

2 a Show that $\sin (x + 45°) = \dfrac{1}{\sqrt{2}}(\sin x + \cos x)$.

   b Hence solve the equation $\sin (x + 45°) = \sqrt{2} \cos x$, for $0° \leqslant x \leqslant 360°$.

3 a Simplify

    i $\sin (A + B) + \sin (A - B)$ **ii** $\cos (A + B) + \cos (A - B)$

   b Hence prove the identity $\dfrac{\sin (A + B) + \sin (A - B)}{\cos (A + B) + \cos (A - B)} \equiv \tan A$

4 Find the values of θ, where $0° \leq \theta \leq 360°$, such that

   a $\cos 2\theta = 1 + \sin \theta$ **b** $\sin 2\theta = \cot \theta$

5 Find the values of x, where $0 \leq x \leq 2\pi$, such that

   a $\cos 2x = \cos x$ **b** $\cos x = \cos \frac{1}{2}x$

6 **a** Prove the identity $\tan A + \cot A \equiv 2\operatorname{cosec} 2A$.

   b Hence, for $0 < x < 2\pi$, solve the equation $\tan x + \cot x = 8$, giving your answer in radians correct to two decimal places.

7 **a** Express $\cos 2A$ in terms of
     i $\cos A$ **ii** $\sin A$.

   b Prove that $\dfrac{2\cos A - \sec A}{\operatorname{cosec} A - 2\sin A} \equiv \tan A$.

8 **a** Expand $\sin(X - Y)$.

   b By letting $X = 4A$ and $Y = 2A$, or otherwise, prove the identity

$$\frac{\sin 4A \cos 2A - \cos 4A \sin 2A}{\sin A} \equiv 2\cos A$$

9 Given that $\tan(A + B) = 1$ and $\tan A = \frac{1}{3}$, find the value of $\tan B$.

10 **a** Prove by a counter example that the statement

      '$\cot(A + B) \equiv \cot A + \cot B$ for all A and B'

    is false.

   b Prove that $\cot(A + B) \equiv \dfrac{\cot A \cot B - 1}{\cot A + \cot B}$.

SKILLS CHECK **2B EXTRA** is on the CD

2.5 Expressions for $a\cos\theta + b\sin\theta$

Show familiarity with the expression of $a\cos\theta + b\sin\theta$ in the forms of $R\sin(\theta \pm \alpha)$ or $R\cos(\theta \pm \alpha)$.

The expression $a\cos\theta + b\sin\theta$ can be written in the form $R\sin(\theta \pm \alpha)$ or $R\cos(\theta \pm \alpha)$.

> **Note:**
> This form is useful for solving equations, drawing graphs and finding maximum and minimum values of functions.

Example 2.13 It is given that $f(\theta) = 2\cos\theta + 3\sin\theta$.

   a Express $f(\theta)$ in the form $R\cos(\theta - \alpha)$, where $R > 0$ and $0° < \alpha < 90°$.

   b Hence solve the equation $2\cos\theta + 3\sin\theta = 3$, where $0° < \theta < 360°$. If an answer is not exact, give it correct to one decimal place.

Step 1: Use an appropriate identity to expand the trig expression.

a
$$R\cos(\theta - \alpha) \equiv 2\cos\theta + 3\sin\theta$$
$$\Rightarrow \quad R(\cos\theta\cos\alpha + \sin\theta\sin\alpha) \equiv 2\cos\theta + 3\sin\theta$$
$$R\cos\theta\cos\alpha + R\sin\theta\sin\alpha \equiv 2\cos\theta + 3\sin\theta$$

> **Recall:**
> $\cos(A - B)$
> $\equiv \cos A\cos B + \sin A\sin B$

Step 2: Equate coefficients to form two equations that enable you to work out R and α.

Equating coefficients of $\cos\theta$:
$$R\cos\alpha = 2 \qquad \qquad ①$$
Equating coefficients of $\sin\theta$:
$$R\sin\alpha = 3 \qquad \qquad ②$$

Step 3: Divide the equations to find α.

Equation ② ÷ equation ① gives
$$\frac{R\sin\alpha}{R\cos\alpha} = \frac{3}{2}$$
$$\Rightarrow \quad \tan\alpha = \tfrac{3}{2}$$
$$\alpha = 56.30\ldots°$$

> **Recall:**
> $\dfrac{\sin\alpha}{\cos\alpha} = \tan\alpha$

> **Tip:**
> $\alpha = \tan^{-1}\left(\frac{b}{a}\right) = \tan^{-1}\left(\frac{3}{2}\right)$

Step 4: Square and add the equations to find R.

Squaring ① and ② and adding gives
$$R^2\cos^2\alpha + R^2\sin^2\alpha = 2^2 + 3^2$$
$$R^2(\cos^2\alpha + \sin^2\alpha) = 13$$
$$R^2 = 13$$
$$R = \sqrt{13}$$

> **Recall:**
> $\cos^2\alpha + \sin^2\alpha = 1$

> **Tip:**
> $R = \sqrt{a^2 + b^2} = \sqrt{2^2 + 3^2}$

Step 5: Write $f(\theta)$ in the required format.

So $\;f(\theta) = 2\cos\theta + 3\sin\theta \equiv \sqrt{13}\cos(\theta - 56.30\ldots°)$

Step 6: Use the format found in **a** to form a simple trig equation and solve.

b
$$2\cos\theta + 3\sin\theta = 3$$
$$\Rightarrow \sqrt{13}\cos(\theta - 56.30\ldots°) = 3$$
$$\Rightarrow \quad \cos(\theta - 56.30\ldots°) = \frac{3}{\sqrt{13}} = 0.8320\ldots$$

Let $x = \theta - 56.30\ldots°$.
The equation becomes $\cos x = 0.8320\ldots$.
Now $\qquad 0° < \theta < 360°$
so $\quad -56.30° < x < 303.69°$
PV of $x = 33.69\ldots°$

> **Tip:**
> Consider the range required for x. In this case you will need to consider negative values.

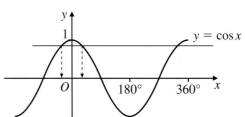

> **Note:**
> The next positive value is $360° - 33.69\ldots° = 326.30\ldots°$ which is out of range.

The other value in range is $-33.69\ldots°$.
So $\theta - 56.30\ldots° = -33.69\ldots°,\ 33.69\ldots°$
$$\theta = 22.6°\ (1\ \text{d.p.}),\ 90°$$

Example 2.14 It is given that $f(x) = 3\sin x - 3\cos x$.

a Find exact values of R and α such that $f(x) \equiv R\sin(x - \alpha)$, where $R > 0$ and $0 < \alpha < \tfrac{1}{2}\pi$.

b The diagram shows a sketch of $y = f(x)$ for $0 \leqslant x \leqslant 2\pi$. It crosses the y-axis at A and the x-axis at B and C. It has a maximum point at D and a minimum point at E.

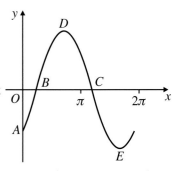

  i Describe the transformations that map the curve $y = \sin x$ onto the curve $y = f(x)$.

  ii Find the exact coordinates of points A, B, C, D and E.

Step 1: Use an appropriate identity to expand the trig expression.

a

$$R \sin (x - \alpha) \equiv 3 \sin x - 3 \cos x$$

$$\Rightarrow \quad R(\sin x \cos \alpha - \cos x \sin \alpha) \equiv 3 \sin x - 3 \cos x$$

$$R \sin x \cos \alpha - R \cos x \sin \alpha \equiv 3 \sin x - 3 \cos x$$

Recall:
$$\sin(A - B)$$
$$\equiv \sin A \cos B - \cos A \sin B$$

Step 2: Equate coefficients to form two equations that enable you to work out R and α.

Equating coefficients of $\sin x$

$$R \cos \alpha = 3 \qquad \qquad ①$$

Equating coefficients of $\cos x$

$$R \sin \alpha = 3 \qquad \qquad ②$$

Equation ② ÷ equation ① gives

$$\frac{R \sin \alpha}{R \cos \alpha} = \frac{3}{3}$$

$$\Rightarrow \qquad \tan \alpha = 1$$

$$\alpha = \tfrac{1}{4}\pi$$

Recall:
$$\frac{\sin \alpha}{\cos \alpha} \equiv \tan \alpha$$

Tip:
Exact values are required, so write the angle in terms of π.

Squaring ① and ② and adding gives

$$R^2 \cos^2 \alpha + R^2 \sin^2 \alpha = 3^2 + 3^2$$

$$R^2 (\cos^2 \alpha + \sin^2 \alpha) = 18$$

$$R^2 = 18$$

$$R = \sqrt{18} = \sqrt{9 \times 2} = 3\sqrt{2}$$

So $f(x) = 3\sqrt{2} \sin(x - \tfrac{1}{4}\pi)$, with $R = 3\sqrt{2}$ and $\alpha = \tfrac{1}{4}\pi$.

Recall:
$$\cos^2 \alpha + \sin^2 \alpha \equiv 1$$

Tip:
Leave your answer in surd form. It is good practice to simplify it.

Step 3: Describe appropriate transformations for $y = af(x - b)$.
Step 4: Identify the coordinates of the intercepts with the axes by setting $x = 0$ and $y = 0$.

b i To transform the curve $y = \sin x$ to $y = 3\sqrt{2} \sin (x - \tfrac{1}{4}\pi)$, translate by $\tfrac{1}{4}\pi$ units in the positive x-direction and stretch by $3\sqrt{2}$ in the y-direction.

Recall:
Transformations (Section 1.4).

ii $y = 3\sqrt{2} \sin (x - \tfrac{1}{4}\pi)$

When $x = 0$, $y = 3\sqrt{2} \sin \left(-\tfrac{1}{4}\pi\right) = 3\sqrt{2} \times \left(-\tfrac{1}{\sqrt{2}}\right) = -3$

So the coordinates of A are $(0, -3)$.

When $y = 0$, $3\sqrt{2} \sin \left(x - \tfrac{1}{4}\pi\right) = 0$

$$\Rightarrow \qquad x - \tfrac{1}{4}\pi = 0 \qquad \text{or } x - \tfrac{1}{4}\pi = \pi$$

$$\Rightarrow \qquad x = \tfrac{1}{4}\pi \qquad \qquad x = \tfrac{5}{4}\pi$$

So B is the point $\left(\tfrac{1}{4}\pi, 0\right)$ and C is the point $\left(\tfrac{5}{4}\pi, 0\right)$.

Recall:
$$\sin 0 = \sin \pi = 0$$

Step 5: Identify the coordinates of the maximum and minimum points by considering the range of the transformed curve.

The maximum value of $\sin x$ is 1 and it occurs when $x = \tfrac{1}{2}\pi$.

Hence the maximum value of $3\sqrt{2} \sin (x - \tfrac{1}{4}\pi)$ is $3\sqrt{2} \times 1 = 3\sqrt{2}$, and it occurs when $x - \tfrac{1}{4}\pi = \tfrac{1}{2}\pi \Rightarrow x = \tfrac{3}{4}\pi$.

So D is the point $\left(\tfrac{3}{4}\pi, 3\sqrt{2}\right)$.

Recall:
$$\sin \tfrac{1}{2}\pi = 1$$

The minimum value of $\sin x$ is -1 and it occurs when $x = \tfrac{3}{2}\pi$.

Hence the minimum value of $3\sqrt{2} \sin \left(x - \tfrac{1}{4}\pi\right)$ is

$3\sqrt{2} \times (-1) = -3\sqrt{2}$, and it occurs when $x - \tfrac{1}{4}\pi = \tfrac{3}{2}\pi \Rightarrow x = \tfrac{7}{4}\pi$.

So E is the point $\left(\tfrac{7}{4}\pi, -3\sqrt{2}\right)$.

Recall:
$$\sin \tfrac{3}{2}\pi = -1$$

Which alternative format is preferable?

If the alternative format of $a \cos \theta + b \sin \theta$ is not specified in the question, it is advisable to ensure that α is acute, that is $0° < \alpha < 90°$ or $0 < \alpha < \tfrac{1}{2}\pi$.

For example, you could use the following alternatives:

For $\quad 3 \cos \theta - 4 \sin \theta \quad$ use $R \cos (\theta + \alpha)$

For $\quad 4 \sin \theta - 3 \cos \theta \quad$ use $R \sin (\theta - \alpha)$

Tip:
In the cos format, the sign in the middle is opposite to the sign in the expression. In the sin format, the sign in the middle is the same as the sign in the expression.

However for $3 \cos \theta + 4 \sin \theta$ you could use either format:
$$3 \cos \theta + 4 \sin \theta \equiv R \cos (\theta - \alpha)$$
or $\qquad 4 \sin \theta + 3 \cos \theta \equiv R \sin (\theta + \alpha)$

SKILLS CHECK **2C: Expressions for** $a \cos \theta + b \sin \theta$

1 a Express $2 \cos x + \sin x$ in the form $R \cos (x - \alpha)$, where $R > 0$ and $0° < \alpha < 90°$.

  b Hence solve the equation $2 \cos x + \sin x = 1$, for values of x in the interval $0° < x < 360°$.

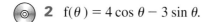

2 $f(\theta) = 4 \cos \theta - 3 \sin \theta$.

  a Express $f(\theta)$ in the form $R \cos (\theta + \alpha)$, where $R > 0$ and $0° < \alpha < 90°$, giving the value of α in degrees, correct to one decimal place.

  b Hence

    i write down the maximum value of $f(\theta)$,

    ii find the largest negative value of θ at which $f(\theta)$ is maximum.

3 a Express $\sin x + \cos x$ in the form $R \sin(x + \alpha)$, where $R > 0$ and $0 < \alpha < \frac{1}{2}\pi$, giving the exact value of α.

  b Hence show that one of the solutions of the equation
$$\sin x + \cos x = \frac{1}{\sqrt{2}} \text{ is } x = \tfrac{7}{12}\pi \text{ and find the exact value of the other solution in the interval}$$
$-\pi < x < \pi$.

4 a Express $3 \sin x - 4 \cos x$ in the form $R \sin (x - \alpha)$, where $R > 0$ and $0° < \alpha < 90°$.

  b Describe how the graph of $y = 3 \sin x - 4 \cos x$ can be obtained from the graph of $y = \sin x$ by applying appropriate transformations.

5 The expression $k \cos x + 15 \sin x$ can be written in the form $17 \cos (x - \alpha)$, where $0° < \alpha < 90°$ and $k > 0$. Find the values of α and k.

SKILLS CHECK **2C EXTRA** is on the CD

Examination practice 2   Trigonometry

1 Solve
$$6 \cos x = 1 + \sec x$$
for $0° < x < 360°$, giving answers correct to one decimal place where necessary.

2 Solve, showing clear working and giving your answers correct to the nearest degree, where appropriate,
$$3 \sec^2 x + \tan x - 5 = 0,$$
for $0° \leqslant x \leqslant 360°$.

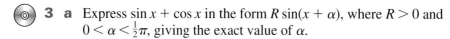

3 Solve, giving your answers in terms of π,
$$\cos 2x = 3 \cos x - 2, \quad 0 \leqslant x \leqslant 2\pi.$$

4 Without using a calculator, find the exact value of
$$\left(\sin 22\tfrac{1}{2}° + \cos 22\tfrac{1}{2}° \right)^2.$$

[OCR Jan 2002]

5 Given that $x = \cos^{-1}\left(\frac{1}{3}\right)$, find the exact value of $\cos\left(\frac{2}{3}\pi - x\right)$ in a simplified form.   [OCR June 2000]

 6 Prove that
$$\frac{1 + \tan^2 \theta}{1 - \tan^2 \theta} \equiv \sec 2\theta$$

7 a Express $7 \cos \theta + 24 \sin \theta$ in the form $R \cos (\theta - \alpha)$, where $R > 0$ and $0 < \alpha < \dfrac{\pi}{2}$.

  b Hence, or otherwise, solve the equation

      $7 \cos \theta + 24 \sin \theta = 12.5$,

    for $0 < \theta < 2\pi$, giving your answers to one decimal place.

 8 a Express $6 \sin x \cos x + 4 \sin^2 x - 2$ in the form $a \sin 2x + b \cos 2x$, where a and b are constants to be found.

  b Hence solve

      $6 \sin x \cos x + 4 \sin^2 x - 2 = 0$

    for $0° < x < 360°$.

9 i Express $\sec^2 \theta + \csc^2 \theta$ in terms of $\sin \theta$ and $\cos \theta$, giving your answer as a single fraction as simply as possible.

  ii Hence prove that
    $\sec^2 \theta + \csc^2 \theta = 4 \csc^2 2\theta$.

  iii Find the values of θ, for $0 < \theta < \frac{1}{2}\pi$, such that
    $\sec^2 \theta + \csc^2 \theta = 10$.   [OCR June 2003]

10 Find all the solutions, in the interval $0 < x < 2\pi$, for which

    $\sin \left(x - \frac{1}{6}\pi\right) = \sqrt{3} \cos \left(x + \frac{1}{6}\pi\right)$,

giving your answers in radians, correct to two decimal places.

11 i Write down the formula for $\cos 2x$ in terms of $\cos x$.

  ii Prove the identity $\dfrac{4 \cos 2x}{1 + \cos 2x} \equiv 4 - 2 \sec^2 x$.

  iii Solve, for $0 < x < 2\pi$, the equation $\dfrac{4 \cos 2x}{1 + \cos 2x} = 3 \tan x - 7$.   [OCR June 2005]

12 i Express $3 \sin \theta + 2 \cos \theta$ in the form $R \sin (\theta + \alpha)$, where $R > 0$ and $0° < \alpha < 90°$.

  ii Hence solve the equation $3 \sin \theta + 2 \cos \theta = \frac{7}{2}$, giving all the solutions for which $0° < \theta < 360°$.
    [OCR June 2005]

3 Differentiation and integration

3.1 Differentiation of e^x and $\ln x$

Use the derivatives of e^x and $\ln x$, together with constant multiples, sums and differences.

In *Core 1* and *Core 2* you learnt how to differentiate powers of x, where

$$y = ax^n \Rightarrow \frac{dy}{dx} = nax^{n-1}$$

You also used the fact that

$$y = f(x) \pm g(x) \Rightarrow \frac{dy}{dx} = f'(x) \pm g'(x)$$

In *Core 3* you also need the following results:

- $y = e^x \Rightarrow \dfrac{dy}{dx} = e^x$

- $y = \ln x \Rightarrow \dfrac{dy}{dx} = \dfrac{1}{x}$

Recall:
Multiply by the power of x and reduce the power by 1 (C1 Section 4.2).

Recall:
You can differentiate term by term.

Example 3.1 Differentiate with respect to x:

    a $y = x^3 - 2e^x$ **b** $t = \ln(4x^3)$

Step 1: Differentiate term by term.

  a $y = x^3 - 2e^x$

       $\dfrac{dy}{dx} = 3x^2 - 2e^x$

Step 1: Simplify the expression using log laws.

Step 2: Differentiate term by term.

  b $t = \ln(4x^3)$

        $= \ln 4 + \ln x^3$

        $= \ln 4 + 3 \ln x$

       $\dfrac{dt}{dx} = 0 + 3 \times \dfrac{1}{x}$

           $= \dfrac{3}{x}$

Tip:
Where possible, simplify log terms before differentiating.

Note:
$\ln(4x^3) \neq 3\ln(4x)$

Recall:
$\log(xy) = \log x + \log y$
$\log x^a = a \log x$
(C2 Section 3.5).

Note:
$\ln 4$ is a constant, so its derivative is 0.

Example 3.2 A curve C, with equation $y = x^2 + 3e^x - 1$, crosses the y-axis at the point P. Find an equation of the normal to C at P, giving your answer in the form $ax + by = c$.

Step 1: Find $\dfrac{dy}{dx}$.

   $y = x^2 + 3e^x - 1$

   $\dfrac{dy}{dx} = 2x + 3e^x$

Step 2: Substitute the x-value to get the gradient of the tangent at P.

When $x = 0$, $\dfrac{dy}{dx} = 2(0) + 3e^0 = 3$

The gradient of the tangent at P is 3

Step 3: Find the gradient of the normal at P.

$\Rightarrow$ gradient of normal at P is $-\frac{1}{3}$

Tip:
Be careful when evaluating $3e^0$. It is 3, not 0 or 1.

Recall:
The tangent and normal are perpendicular, so the product of their gradients is -1 (C1 Section 3.3).

Step 4:Substitute x into the equation of the curve to get y.	When $x = 0$, $y = 0^2 + 3e^0 - 1 = 3 - 1 = 2$	**Recall:** Equation of line $y - y_1 = m(x - x_1)$ (C1 Section 3.2).

Step 5: Use an appropriate straight line equation.

Equation of the normal at $P(0, 2)$:

$$y - 2 = -\tfrac{1}{3}(x - 0)$$
$$y - 2 = -\tfrac{1}{3}x$$
$$3y - 6 = -x$$
$$3y = 6 - x$$

Step 6: Rearrange into the required format.

$$x + 3y = 6$$

Note: If you are not asked to give a specific format, leave the equation in a convenient form.

3.2 The chain rule

Differentiate composite functions using the chain rule.

The **chain rule** is one of the most useful results in differentiation. It enables composite functions such as $(5x - 3)^7$, $\tfrac{1}{2}e^{3x + 5}$ and $\ln \dfrac{1}{\sqrt{2x + 9}}$ to be differentiated.

Recall: A composite function is formed by combining two or more functions (Section 1.2).

If y is a function of t and t is a function of x, then, by the chain rule,

$$\frac{dy}{dx} = \frac{dy}{dt} \times \frac{dt}{dx}$$

An alternative version of the chain rule, using function notation, is

$$\frac{d}{dx} f(g(x)) = f'(g(x)) \times g'(x)$$

Note: Sometimes this is known as 'differentiating a function of a function'.

Example 3.3 Find $\dfrac{dy}{dx}$ when

a $y = (5x - 3)^7$ **b** $y = \tfrac{1}{2}e^{3x + 5}$ **c** $y = \ln \dfrac{1}{\sqrt{2x + 9}}$

a $y = (5x - 3)^7$

Step 1: Define t as a function of x; y is now a function of t.

Step 2: Differentiate t with respect to x, and y with respect to t.

Step 3: Rewrite t in terms of x.

Step 4: Apply the chain rule.

Let $t = 5x - 3$, then $y = t^7$

So $\dfrac{dt}{dx} = 5$

and $\dfrac{dy}{dt} = 7t^6$

$\qquad = 7(5x - 3)^6$

$\dfrac{dy}{dx} = \dfrac{dy}{dt} \times \dfrac{dt}{dx}$

$\qquad = 7(5x - 3)^6 \times 5$

$\qquad = 35(5x - 3)^6$

b $y = \tfrac{1}{2}e^{3x + 5}$

Step 1: Define t as a function of x; y is now a function of t.

Step 2: Differentiate t with respect to x, and y with respect to t.

Step 3: Rewrite t in terms of x.

Step 4: Apply the chain rule.

Let $t = 3x + 5$, then $y = \tfrac{1}{2}e^t$

So $\dfrac{dt}{dx} = 3$

and $\dfrac{dy}{dt} = \tfrac{1}{2}e^t$

$\qquad = \tfrac{1}{2}e^{3x + 5}$

$\dfrac{dy}{dx} = \dfrac{dy}{dt} \times \dfrac{dt}{dx}$

$\qquad = \tfrac{1}{2}e^{3x + 5} \times 3 = \tfrac{3}{2}e^{3x + 5}$

Recall: $\dfrac{d}{dx}e^x = e^x$

Step 1: Simplify the expression using the log laws.
Step 2: Define t as a function of x; y is now a function of t.
Step 3: Differentiate t with respect to x, and y with respect to t.
Step 4: Rewrite t in terms of x.
Step 5: Apply the chain rule.

c $y = \ln \dfrac{1}{\sqrt{2x+9}}$

$\quad = \ln (2x+9)^{-\frac{1}{2}}$

$\quad = -\frac{1}{2} \ln (2x+9)$

Let $t = 2x + 9$, then $y = -\frac{1}{2}\ln t$

So $\dfrac{dt}{dx} = 2$

and $\dfrac{dy}{dt} = -\dfrac{1}{2} \times \dfrac{1}{t}$

$\quad = -\dfrac{1}{2(2x+9)}$

$\dfrac{dy}{dx} = \dfrac{dy}{dt} \times \dfrac{dt}{dx}$

$\quad = -\dfrac{1}{2(2x+9)} \times 2$

$\quad = -\dfrac{1}{(2x+9)}$

Example 3.4 The point $P(1, -1)$ lies on the curve with equation $y = \dfrac{1}{(2x-3)^5}$. Find the gradient of the curve at P.

Step 1: Write in index form
Step 2: Define t as a function of x; y is now a function of t.
Step 3: Differentiate t with respect to x, and y with respect to t.
Step 4: Rewrite t in terms of x.

$y = \dfrac{1}{(2x-3)^5} = (2x-3)^{-5}$

Let $t = 2x - 3$, then $y = t^{-5}$

So $\dfrac{dt}{dx} = 2$

and $\dfrac{dy}{dt} = -5t^{-6}$

$\quad = -\dfrac{5}{(2x-3)^6}$

Step 5: Apply the chain rule.

$\dfrac{dy}{dx} = \dfrac{dy}{dt} \times \dfrac{dt}{dx}$

$\quad = -\dfrac{5}{(2x-3)^6} \times 2$

$\quad = -\dfrac{10}{(2x-3)^6}$

Step 6: Substitute the given value and evaluate.

When $x = 1$, $\dfrac{dy}{dx} = -\dfrac{10}{(-1)^6} = -10$

So the gradient of the curve at P is -10.

Some standard derivatives using the chain rule

The following results come from the chain rule and are useful to remember.

$$y = e^{kx} \Rightarrow \dfrac{dy}{dx} = ke^{kx}$$

$$y = e^{ax+b} \Rightarrow \dfrac{dy}{dx} = ae^{ax+b}$$

$$y = \ln(kx) \Rightarrow \frac{dy}{dx} = \frac{1}{x}$$

$$y = \ln(ax + b) \Rightarrow \frac{dy}{dx} = \frac{a}{ax + b}$$

$$y = (ax + b)^n \Rightarrow \frac{dy}{dx} = an(ax + b)^{n-1}$$

Example 3.5 The curve C, with equation $y = 2e^{2x} - 4x$ has a stationary point at P. Find the coordinates of P and determine the nature of the point.

Step 1: Differentiate.

$$y = 2e^{2x} - 4x$$
$$\frac{dy}{dx} = 2 \times 2e^{2x} - 4 = 4e^{2x} - 4$$

Recall:
$$\frac{d}{dx} e^{kx} = ke^{kx}$$

Step 2: Put $\frac{dy}{dx} = 0$ and solve for x.

At P, $\dfrac{dy}{dx} = 0$

so $4e^{2x} - 4 = 0$
$$4(e^{2x} - 1) = 0$$
$$e^{2x} - 1 = 0$$
$$e^{2x} = 1$$

Recall:
At a stationary point the gradient is zero. (C1 Section 4.6).

Taking natural logs of both sides gives

$$\ln e^{2x} = \ln 1$$
$$\Rightarrow \quad 2x \ln e = 0$$
$$2x = 0$$
$$x = 0$$

Recall:
$\ln x$ is the inverse of e^x; $\ln 1 = 0$; $\ln e = 1$ (Section 1.6).

Step 3: Substitute the x-value into the equation of the curve to find y.

When $x = 0$, $y = 2e^{2(0)} - 4 \times 0 = 2 \times 1 - 0 = 2$, so P is the point $(0, 2)$.

Recall:
$e^0 = 1$ (Section 1.6).

Step 4: Differentiate $\frac{dy}{dx}$ to find $\frac{d^2y}{dx^2}$.

$$\frac{d^2y}{dx^2} = 4 \times 2e^{2x} = 8e^{2x}$$

At P, $x = 0$, so $\dfrac{d^2y}{dx^2} = 8e^{2(0)} = 8 \times 1 = 8$.

Step 5: Determine the sign of $\frac{d^2y}{dx^2}$ at the stationary point.

Since $\dfrac{d^2y}{dx^2} > 0$, P is a minimum point.

Note:
If $\frac{d^2y}{dx^2} = 0$, then use the alternative method of testing the sign of $\frac{dy}{dx}$ either side of the stationary point.

Example 3.6 Given that $A = \ln(5x + 2) + 6x$, find

a $\dfrac{dA}{dx}$

b the value of $\dfrac{d^2A}{dx^2}$ when $x = 0$

Step 1: Differentiate term by term.

a $A = \ln(5x + 2) + 6x$
$$\frac{dA}{dx} = \frac{5}{5x + 2} + 6$$

Recall:
$$\frac{d}{dx} \ln(ax + b) = \frac{a}{ax + b}$$

Step 2: Write terms in index form and differentiate again.

b $\dfrac{dA}{dx} = 5(5x + 2)^{-1} + 6$

$$\frac{d^2A}{dx^2} = 5(-1)(5)(5x + 2)^{-2} + 0 = -\frac{25}{(5x + 2)^2}$$

Tip:
In parts **a** and **b**, you could use the chain rule with $t = 5x + 2$.

Tip:
The derivative of a constant is zero.

Step 3: Substitute the given value of x.

When $x = 0$, $\dfrac{d^2A}{dx^2} = -\dfrac{25}{(5 \times 0 + 2)^2} = -\dfrac{25}{4}$

You may be asked to find rates of change, for example in exponential growth and decay questions.

Recall:
Exponential growth and decay (Section 1.7).

If $y = f(t)$, where t is time, then the **rate of change** of y with t is given by $f'(t)$ or $\dfrac{dy}{dt}$.

If $\dfrac{dy}{dt} > 0$, y is *increasing* as t increases.

If $\dfrac{dy}{dt} < 0$, y is *decreasing* as t increases.

The rate of change at a given value of t is the gradient of the tangent to the curve $y = f(t)$ at that point.

Note:
If y is increasing, the tangent has a positive gradient and if y is decreasing, the tangent has a negative gradient.

Example 3.7 The mass, in grams, of a radioactive substance after t years, where $t \geqslant 0$, is given by

$$m = 50e^{-kt}$$

where k is a positive constant.

a What is the value of m when $t = 0$?

After 300 years, the mass will be half its value when $t = 0$.

b Find the value of k, correct to two significant figures.

c Find the rate at which the mass is decreasing when $t = 250$.

Step 1: Substitute $t = 0$.

a $\qquad\qquad m = 50e^{-kt}$

When $t = 0$, $m = 50e^0$

$\qquad\qquad\qquad = 50$

When $t = 0$, the mass is 50 grams.

Recall:
$e^0 = 1$

Note:
This is sometimes called the initial mass.

Step 2: Use the given information and substitute $t = 300$.

b When $t = 300$, $m = 0.5 \times 50 = 25$

So $\quad 25 = 50e^{-k \times 300}$

$e^{-300k} = 0.5$

$-300k = \ln 0.5$

Tip:
Divide by 50 first then write the equation in log format.

Step 3: Solve for k.

$k = \dfrac{\ln 0.5}{-300} = 0.002310\ldots = 0.0023$ (2 s.f.)

Step 4: Differentiate m with respect to t.

c $m = 50e^{-kt}$

$\dfrac{dm}{dt} = -50ke^{-kt}$

Note:
The rate of change of the mass (with time) is given by $\dfrac{dm}{dt}$.

Step 5: Substitute known values and evaluate.

When $t = 250$, $\dfrac{dm}{dt} = -50(0.00231\ldots)e^{-0.00231\ldots \times 250}$

$\qquad\qquad\qquad = -0.06483\ldots$

So, when $t = 250$, the mass is decreasing at a rate of 0.065 grams/year (2 s.f.).

Tip:
The negative value indicates that the mass is decreasing.

SKILLS CHECK **3A: Differentiation of functions**

1 Differentiate with respect to x:

a $5e^x$ **b** $\ln x^3$ **c** $2x^3 - e^x$ **d** $\ln \dfrac{5}{x^2}$

2 Differentiate with respect to x:

a $2e^{-3x}$ **b** $4\ln(3x)$ **c** $3e^{x^2 - 6x}$ **d** $4\ln(5 - 2x)$

3 a Simplify $\dfrac{\ln 5x}{\sqrt[3]{2x + 7}}$.

 b Given that $y = \ln \dfrac{5x}{\sqrt[3]{2x + 7}}$, find $\dfrac{dy}{dx}$.

4 Differentiate with respect to x:

 a $y = (5x + 6)^4$

 b $z = \sqrt{(2x - 1)}$

 c $t = \dfrac{1}{x + 2}$

 5 Find an equation of the tangent to the curve $y - (\frac{1}{2}x^2 - 1)^3$ at the point where $x - -2$.

6 The curve C has equation $y = x^2 - 8 \ln x$, $x > 0$. Show that C has only one stationary point and find its coordinates.

 7 A curve C has equation $y = 3e^x - 2 \ln 5x$.

 a Find $\dfrac{dy}{dx}$.

 b Find an equation of the tangent to the curve at $x = 1$.

 c Find the exact y-coordinate of the point where this tangent crosses the y-axis. Give your answer in the form $a - \ln b$.

8 It is given that $f(x) = (e^x + 3)(e^{2x} - 5)$.

 a Expand the brackets.

 b Find $f'(x)$.

 c Find $f''(0)$.

 9 Find the gradient of the curve $y = \dfrac{e^{3x} + 1}{e^{2x}}$ when $x = 0$.

10 Differentiate with respect to t:

 a $x = \dfrac{3}{1 - t}$

 b $y = \ln (2t + 3)^4$

11 Is the function $f(x) = e^x - x^2$ increasing or decreasing when $x = 2$? Show sufficient working to support your answer.

12 The number of bacteria present in a culture at a time t hours after the beginning of an experiment is given by $y = Ae^{2t}$.

 a Find the number of bacteria after one hour if there are 1000 bacteria at the beginning of the experiment.

 b Sketch the graph of y against t showing the coordinates of any intercepts with the axes.

 c i Find the rate at which the number of bacteria is increasing after one hour.

 ii Explain how this can be illustrated on the sketch.

SKILLS CHECK **3A EXTRA** is on the CD

Differentiate products and quotients.

The product rule

To differentiate a product of two functions (i.e. an expression formed by multiplying two functions together) use the **product rule**:

If $y = uv$, where u and v are functions of x, then

$$\frac{dy}{dx} = u\frac{dv}{dx} + v\frac{du}{dx}$$

An alternative version of the rule, using function notation, is

$$\frac{d}{dx}(f(x)g(x)) = f'(x)g(x) + f(x)g'(x)$$

Note:
In words: first function $\times$ derivative of second + second function $\times$ derivative of first.

Note:
If you use the alternative version of the rule your working will be in a different order, but the answer will be the same.

Example 3.8 Differentiate with respect to x, simplifying your answers:

    a $y = (3x^2 - 2)(1 - 4x^3)$ **b** $y = 2x^4 e^{3x}$

Step 1: Define u and v in terms of x.

a $y = (3x^2 - 2)(1 - 4x^3)$

Let $u = 3x^2 - 2$, $v = 1 - 4x^3$

Step 2: Differentiate u and v with respect to x.

So $\dfrac{du}{dx} = 6x$ and $\dfrac{dv}{dx} = -12x^2$

Note:
The derivative could be found by multiplying out the brackets and then differentiating without using the product rule.

Step 3: Apply the product rule.

$$\frac{dy}{dx} = u\frac{dv}{dx} + v\frac{du}{dx}$$
$$= (3x^2 - 2) \times (-12x^2) + (1 - 4x^3) \times 6x$$

Step 4: Simplify.

$$= -12x^2(3x^2 - 2) + 6x(1 - 4x^3)$$
$$= 6x[-2x(3x^2 - 2) + (1 - 4x^3)]$$
$$= 6x(-6x^3 + 4x + 1 - 4x^3)$$
$$= 6x(1 + 4x - 10x^3)$$

Tip:
It's good practice to tidy up expressions that result from the product rule even if it is not specifically requested in the question.

b $y = 2x^4 e^{3x}$

Step 1: Define u and v in terms of x.

Let $u = 2x^4$, $v = e^{3x}$

Step 2: Differentiate u and v with respect to x.

So $\dfrac{du}{dx} = 8x^3$ and $\dfrac{dv}{dx} = 3e^{3x}$

Step 3: Apply the product rule.

$$\frac{dy}{dx} = u\frac{dv}{dx} + v\frac{du}{dx}$$
$$= 2x^4 \times 3e^{3x} + e^{3x} \times 8x^3$$

Step 4: Simplify.

$$= 2x^3 e^{3x}(3x + 4)$$

Tip:
Take out common factors.

Example 3.9 The curve C has equation $y = xe^{-x}$. Show that the curve has a stationary point when $x = 1$.

Step 1: Find $\dfrac{dy}{dx}$ by applying the product rule.

$y = xe^{-x}$

Let $u = x$, $v = e^{-x}$

So $\dfrac{du}{dx} = 1$ and $\dfrac{dv}{dx} = -e^{-x}$

$$\frac{dy}{dx} = u\frac{dv}{dx} + v\frac{du}{dx}$$
$$= x \times -e^{-x} + e^{-x} \times 1$$
$$= e^{-x}(-x + 1)$$

Step 2: Substitute the given value and comment.
When $x = 1$, $\dfrac{dy}{dx} = e^{-1}(-1 + 1) = 0$

Since $\dfrac{dy}{dx} = 0$, there is a stationary point when $x = 1$.

> **Note:**
> The question doesn't ask for the coordinates of the stationary point so you are not required to find the y-coordinate.

The quotient rule

To differentiate a quotient of two functions (i.e. an expression formed by dividing one function by another) use the **quotient rule**.

If $y = \dfrac{u}{v}$, where u and v are functions of x, then

$$\frac{dy}{dx} = \frac{v\dfrac{du}{dx} - u\dfrac{dv}{dx}}{v^2}$$

An alternative version of the rule, using function notation, is

$$\frac{d}{dx}\left(\frac{f(x)}{g(x)}\right) = \frac{f'(x)g(x) - f(x)g'(x)}{(g(x))^2}$$

> **Note:**
> In words: bottom function × derivative of top − top function × derivative of bottom, all over bottom function squared.

> **Note:**
> If you use the alternative version of the rule your working will be in a different order, but the answer will be the same.

Example 3.10 Differentiate $\dfrac{e^{3x}}{x^2}$ with respect to x, simplifying your answer.

Let $y = \dfrac{e^{3x}}{x^2}$

Step 1: Define u and v in terms of x.
Let $u = e^{3x}$, $v = x^2$

Step 2: Differentiate u and v with respect to x.
So $\dfrac{du}{dx} = 3e^{3x}$

and $\dfrac{dv}{dx} = 2x$

> **Recall:**
> $\dfrac{d}{dx}(e^{kx}) = ke^{kx}$ (Section 3.2).

Step 3: Apply the quotient rule
$$\frac{dy}{dx} = \frac{v\dfrac{du}{dx} - u\dfrac{dv}{dx}}{v^2}$$
$$= \frac{x^2 \times 3e^{3x} - e^{3x} \times 2x}{(x^2)^2}$$

Step 4: Simplify
$$= \frac{xe^{3x}(3x - 2)}{x^4}$$
$$= \frac{e^{3x}(3x - 2)}{x^3}$$

> **Note:**
> You could differentiate this as a product of e^{3x} and x^{-2} to give
> $$\frac{dy}{dx} = e^{3x}(-2x^{-3}) + x^{-2}(3e^{3x})$$
> which then simplifies to the same result.

Example 3.11 Given that $f(x) = \dfrac{x}{\sqrt{1 + x}}$, $x > -1$,

a show that $f'(x) = \dfrac{2 + x}{2(1 + x)^{\frac{3}{2}}}$

b find the exact value of $f'(1)$, in simplified surd form.

a $f(x) = \dfrac{x}{\sqrt{1+x}}$

Let $u = x$, $v = \sqrt{1+x} = (1+x)^{\frac{1}{2}}$

Step 1: Define u and v in terms of x
Step 2: Differentiate u and v with respect to x.

So $\dfrac{du}{dx} = 1$

and $\dfrac{dv}{dx} = \dfrac{1}{2} \times 1 \times (1+x)^{-\frac{1}{2}} = \dfrac{1}{2}(1+x)^{-\frac{1}{2}}$

> **Tip:**
> You can use the chain rule here, but it is easier to remember that
> $\dfrac{d}{dx}(ax+b)^n = an(ax+b)^{n-1}$

Step 3: Apply the quotient rule.

$f'(x) = \dfrac{v \dfrac{du}{dx} - u \dfrac{dv}{dx}}{v^2}$

$= \dfrac{(1+x)^{\frac{1}{2}} \times 1 - x \times \frac{1}{2}(1+x)^{-\frac{1}{2}}}{(\sqrt{1+x})^2}$

Step 4: Simplify.

$= \dfrac{\sqrt{1+x} - \dfrac{x}{2\sqrt{1+x}}}{1+x}$

> **Tip:**
> Multiply every term in the numerator and in the denominator by $2\sqrt{1+x}$.

$= \dfrac{2(1+x) - x}{2(1+x)\sqrt{1+x}}$

> **Tip:**
> $a\sqrt{a} = a^1 \times a^{\frac{1}{2}} = a^{\frac{3}{2}}$

$= \dfrac{2 + 2x - x}{2(1+x)\sqrt{1+x}}$

$= \dfrac{2+x}{2(1+x)^{\frac{3}{2}}}$

Step 5: Substitute the given value of x.

b $f'(1) = \dfrac{2+1}{2(1+1)^{\frac{3}{2}}}$

> **Recall:**
> $a^{\frac{3}{2}} = (\sqrt{a})^3 = \sqrt{a} \times \sqrt{a} \times \sqrt{a}$
> (C1 Section 1.1).

$= \dfrac{3}{2(\sqrt{2})^3}$

$= \dfrac{3}{4\sqrt{2}}$

> **Recall:**
> Rationalise the denominator (C1 Section 1.2).

$= \dfrac{3\sqrt{2}}{8}$

SKILLS CHECK **3B: Product and quotient rules**

1 Using the product rule, differentiate with respect to x:

  a $x^3(x+4)^2$ **b** $e^{2x}x^4$ **c** $x \ln\sqrt{x+3}$

2 Using the quotient rule, differentiate with respect to x:

  a $\dfrac{3x^2}{x-3}$ **b** $\dfrac{e^{\frac{x}{2}}}{2x^3}$ **c** $\dfrac{\ln(x+1)}{x+1}$

3 Differentiate with respect to x:

  a $(x^2+3)^3(5x-4)^5$ **b** $\dfrac{2e^x - 1}{2e^x + 1}$ **c** $\dfrac{x^2}{e^x}$

4 The curve C has the equation $y = \dfrac{\ln x}{x^2}$. Find the gradient of C at the point where $x = e$.

5 Find the stationary points on the following curves:

   a $y = 2xe^x$

   b $y = x^3 e^{-x}$

6 Given that $f(x) = \dfrac{x^2}{x - 3}$,

   a show that $f'(x) = \dfrac{x^2 - 6x}{(x - 3)^2}$,

   b find $f''(x)$ in its simplest form.

7 Find an equation for the normal to the curve $y = \dfrac{x^2}{1 + x^2}$ at the point where $x = 1$.

8 The curve C has equation $y = 2x^{\frac{1}{2}}e^{-x}$. Find the x-coordinate of the stationary point of the curve.

9 **a** Given that $y = x\sqrt{1 + x}$, show that $\dfrac{dy}{dx} = \dfrac{2 + 3x}{2\sqrt{1 + x}}$.

   b Given that $y = \dfrac{x}{1 + 2x}$, show that $\dfrac{dy}{dx} = \dfrac{1}{(1 + 2x)^2}$.

10 A curve C, with equation $y = \dfrac{\ln x}{x}$ has one stationary point, at the point P. Find the exact coordinates of P.

11 It is given that $y = (\ln x)^4$.

   a Find the value of x when $\dfrac{dy}{dx} = 0$.

   b Find the value of $\dfrac{d^2y}{dx^2}$ when $x = e$.

SKILLS CHECK **3B EXTRA** is on the CD

3.4 Finding the derivative when $x = f(y)$

Understand and use the relation $\dfrac{dy}{dx} = \dfrac{1}{\dfrac{dx}{dy}}$.

Sometimes a function is defined in terms of y instead of x. A variation of the chain rule enables the derivative of such a function to be found, where

$$\frac{dy}{dx} = \frac{1}{\dfrac{dx}{dy}}$$

So, if x is given as a function of y, differentiate x with respect to y to find $\dfrac{dx}{dy}$ and then work out the reciprocal to get $\dfrac{dy}{dx}$.

Note:

It is interesting to note that this rule does not apply to the second derivative:

$$\frac{d^2y}{dx^2} \neq \frac{1}{\dfrac{d^2x}{dy^2}}.$$

Example 3.12 Given that $x = y + y^2$,

a find an expression, in terms of y, for $\dfrac{dy}{dx}$,

b calculate the value of $\dfrac{dy}{dx}$ when $y = 3$.

Step 1: Differentiate x with respect to y.

a $x = y + y^2$

$$\frac{dx}{dy} = 1 + 2y$$

Step 2: Use the reciprocal relationship to find $\dfrac{dy}{dx}$.

$$\frac{dy}{dx} = \frac{1}{\dfrac{dx}{dy}} = \frac{1}{1 + 2y}$$

Step 3: Substitute the given value of y.

b When $y = 3$, $\dfrac{dy}{dx} = \dfrac{1}{1 + 6} = \dfrac{1}{7}$

Example 3.13 It is given that $y = \ln x$.

By first expressing x in terms of y, show that $\dfrac{d}{dx}(\ln x) = \dfrac{1}{x}$.

Step 1: Express x in terms of y.

$y = \ln x$

So $x = e^y$

Step 2: Differentiate x with respect to y.

$$\frac{dx}{dy} = e^y$$

Step 3: Rewrite in terms of x.

$$= x$$

Step 4: Use the reciprocal relationship to find $\dfrac{dy}{dx}$.

$$\frac{dy}{dx} = \frac{1}{\dfrac{dx}{dy}} = \frac{1}{x}$$

i.e. $\dfrac{d}{dx}(\ln x) = \dfrac{1}{x}$

3.5 Connected rates of change

Apply differentiation to connected rates of change.

The chain rule can be applied to **connected rates of change**. This is illustrated in the following example.

Example 3.14 A cuboid has a rectangular base of length $4x$ cm and width x cm. The height of the cuboid is $3x$ cm.

a Given that the width of the base is increasing at a rate of 3 cm/s, find the rate of increase of the area of the base when $x = 2$.

b Given that the volume of the cuboid is increasing at a rate of 18 cm³/s, find the rate at which the width of the base is increasing when $x = 1$.

Step 1: Define the variables, stating the units.
Step 2: Express the given rate of change in calculus form.

a Let A be the area of the base, in cm² and t the time in seconds.

It is given that $\dfrac{dx}{dt} = 3$.

Step 3: Express the area in terms of x, differentiate with respect to x.

Step 4: Use the chain rule to differentiate A with respect to t.

Step 5: Substitute known values.

$$A = 4x^2 \Rightarrow \frac{dA}{dx} = 8x$$

We require the rate of change of A, i.e. $\frac{dA}{dt}$

By the chain rule

$$\frac{dA}{dt} = \frac{dA}{dx} \times \frac{dx}{dt}$$
$$= 8x \times 3$$
$$= 24x$$

When $x = 2$, $\frac{dA}{dt} = 24 \times 2 = 48$

The area is increasing at 48 cm²/s when $x = 2$.

Tip:
The rate at which the area is changing (with time) is given by $\frac{dA}{dt}$.

Tip:
Specify the units in your final answer.

Step 1: Define the variables, stating the units.

Step 2: Express the given rate of change in calculus form.

Step 3: Express V in terms of x, differentiate with respect to x.

Step 4: Use the chain rule to differentiate x with respect to t.

Step 5: Substitute known values.

b Let V be the volume of the cuboid, in cm³ and t the time in seconds.

Volume increasing at 18 cm³/s $\Rightarrow \frac{dV}{dt} = 18$

$$V = 12x^3 \Rightarrow \frac{dV}{dx} = 36x^2$$

We require the rate of change of x, i.e. $\frac{dx}{dt}$.

By the chain rule

$$\frac{dx}{dt} = \frac{dx}{dV} \times \frac{dV}{dt}$$

Since $\frac{dV}{dx} = 36x^2$, $\frac{dx}{dV} = \frac{1}{36x^2}$

So $\frac{dx}{dt} = \frac{1}{36x^2} \times 18 = \frac{1}{2x^2}$

When $x = 1$, $\frac{dx}{dt} = \frac{1}{2 \times 1^2} = 0.5$

The width of the base is increasing at 0.5 cm/s when $x = 1$.

Tip:
The rate at which the volume is changing (with time) is given by $\frac{dV}{dt}$.

Recall:
$\frac{dx}{dV} = \frac{1}{\frac{dV}{dx}}$ (Section 3.4).

Example 3.15 The variables P and Q are related by the expression $P = 5Q^2 + 3$. If Q is decreasing at a rate of 4 units/s, find the rate at which P is decreasing when $Q = 7$.

Step 1: Express the given rate of change in calculus form.

Step 2: Differentiate P with respect to Q.

Step 3: Use the chain rule to differentiate P with respect to t.

Step 4: Substitute known values.

Q is decreasing at a rate of 4 units/s $\Rightarrow \frac{dQ}{dt} = -4$

$$P = 5Q^2 + 3 \Rightarrow \frac{dP}{dQ} = 10Q$$

$$\frac{dP}{dt} = \frac{dP}{dQ} \times \frac{dQ}{dt}$$
$$= 10Q \times (-4)$$
$$= -40Q$$

When $Q = 7$, $\frac{dP}{dt} = -40 \times 7 = -280$

So P is decreasing at a rate of 280 units/s.

Tip:
You need the negative, since Q is decreasing.

Note:
Since $\frac{dP}{dt} < 0$, P is also decreasing.

1 Given that $x = \dfrac{y^2 - 5}{2}$, find $\dfrac{dy}{dx}$ in terms of y.

2 Given that $x = \dfrac{y^2 + y}{y - 1}$, find $\dfrac{dy}{dx}$ in terms of y.

3 Given that $x = y^4$, find $\dfrac{dy}{dx}$ in terms of x.

4 Given the curve $x = \dfrac{y^2 + 2}{6}$,

  a find $\dfrac{dy}{dx}$,

  b find an equation for the normal to the curve at the point where $y = 1$.

5 The curve C has equation $x = \dfrac{y}{e^y}$. Find the gradient of the tangent to C at the point where $y = \ln 3$.

6 Given the curve $x = \dfrac{y^2}{4}$, for $x > 0$, $y > 0$,

  a find equations for the tangents to the curve at $x = 1$ and at $x = 4$,

  b find the coordinates of the point of intersection of these two tangents.

7 Given the curve C, with equation $x = \sqrt{y^2 + 25}$, find an equation for the tangent to C at the point $(13, 12)$.

8 Variables x and y are connected by the equation $y = 3x^2 + 5 \ln x$.

  a Given that x is increasing at the rate of 4 units per second, find the rate of increase in y when $x = 2$.

  b Given that x is decreasing at the rate of 4 units per second, find the rate at which y is changing when $x = 3$, stating whether this is an increase or a decrease.

9 The radius of a circular ink blot is increasing at a rate of 0.3 cm/s. Find the rate at which the area is increasing when the radius is 0.5 cm.

10 The volume of a spherical balloon is decreasing at a rate of 25 cm³/s. Find the rate at which the radius is decreasing at the instant when the radius of the balloon is 2 cm. Give your answer correct to two significant figures.

SKILLS CHECK **3C EXTRA** is on the CD

3.6 Integration of e^x and $\dfrac{1}{x}$

Integrate e^x and $\frac{1}{x}$, together with constant multiples, sums and differences.

In *Core 2* you learnt how to integrate powers of x using the following formula:

$$\int ax^n \, dx = \frac{a}{n + 1} x^{n + 1} + c \quad (n \neq -1)$$

Recall:
Integration (C2 Chapter 4).

Recall:
Increase the power of x by 1 and divide by this new power.

Recall also that $\int (f(x) \pm g(x))\,dx = \int f(x)\,dx \pm \int g(x)\,dx$.

Recall:
You can integrate sums and differences term by term.

In *Core 3* the functions you are required to integrate are extended to the following, together with functions that can be integrated by applying a linear substitution:

- $\int e^x\,dx = e^x + c$

- $\int \dfrac{1}{x}\,dx = \ln|x| + c$

Note:
It is possible to incorporate the constant into the log expression by letting $c = \ln k$. Then
$\int \dfrac{1}{x}\,dx = \ln|x| + \ln k = \ln k|x|$

Take special care with constant factors, for example:

$$\int \frac{5}{x}\,dx = 5\int \frac{1}{x}\,dx = 5\ln|x| + c$$

but $\displaystyle\int \frac{1}{5x}\,dx = \frac{1}{5}\int \frac{1}{x}\,dx = \tfrac{1}{5}\ln|x| + c$

Example 3.16 Find   **a** $\displaystyle\int (x^2 + 2e^x)\,dx$ **b** $\displaystyle\int \left(\frac{2}{x} + \frac{1}{3x^2}\right)dx$

Step 1: Integrate term by term.

a $\displaystyle\int (x^2 + 2e^x)\,dx = \tfrac{1}{3}x^3 + 2e^x + c$

Step 2: Write terms in index form where appropriate, before integrating.

b $\displaystyle\int \left(\frac{2}{x} + \frac{1}{3x^2}\right)dx = \int \left(\frac{2}{x} + \frac{1}{3}x^{-2}\right)dx$

$$= 2\ln|x| + \frac{1}{3} \times \frac{x^{-1}}{-1} + c$$

$$= 2\ln|x| - \frac{1}{3x} + c$$

Note:
$\dfrac{2}{x} = 2x^{-1}$, but this format is not particularly helpful here since the formula for $\int x^n\,dx$ is not applicable when $n = -1$.

Example 3.17 **a** Evaluate $\displaystyle\int_e^{e^2} \frac{1}{2x}\,dx$.

     b Given that $\displaystyle\int_{-4}^{-2} \left(x + \frac{5}{x}\right)dx = p + q\ln 2$, where p and q are integers, find the value of p and the value of q.

Step 1: Integrate.

a $\displaystyle\int_e^{e^2} \frac{1}{2x}\,dx = \frac{1}{2}\int_e^{e^2} \frac{1}{x}\,dx$

Step 2: Substitute the limits and evaluate.

$$= \tfrac{1}{2}\big[\ln|x|\big]_e^{e^2}$$
$$= \tfrac{1}{2}(\ln e^2 - \ln e)$$
$$= \tfrac{1}{2}(2\ln e - \ln e)$$
$$= \tfrac{1}{2}\ln e$$
$$= \tfrac{1}{2}$$

Tip:
Take out the numerical factor.

Recall:
Substitute the upper limit, then subtract the result when the lower limit is substituted.

Recall:
$\ln e = 1$ (Section 1.6).

Step 3: Integrate term by term.

b $\displaystyle\int_{-4}^{-2} \left(x + \frac{5}{x}\right)dx = \big[\tfrac{1}{2}x^2 + 5\ln|x|\big]_{-4}^{-2}$

Step 4: Substitute the limits and evaluate.

$$= \tfrac{1}{2}(-2)^2 + 5\ln|-2| - \big(\tfrac{1}{2}(-4)^2 + 5\ln|-4|\big)$$
$$= 2 + 5\ln 2 - (8 + 5\ln 4)$$

Step 5: Compare with the given result.

$$= -6 + 5\ln\tfrac{2}{4}$$
$$= -6 + 5\ln 2^{-1}$$
$$= -6 - 5\ln 2$$

Tip:
If $a > 0$, $\ln|-a| = \ln a$.

Recall:
Log laws applied to logs to the base e: $\ln a - \ln b = \ln\frac{a}{b}$ and $\ln a^{-1} = -\ln a$.

Hence $p = -6$ and $q = -5$.

Integrate expressions involving a linear substitution, e.g. $(2x - 1)^8$, $e^{3x + 2}$.

Using the substitution u, where u is a function of x, $y = \int f(x)\, dx$
becomes $y = \int f(x)\, \dfrac{dx}{du}\, du$.

Note:
In C3, the substitution will be a linear one, that is, of the form $ax + b$.

Example 3.18 Using the substitution $u = 2x - 1$, find $\int (2x - 1)^8\, dx$.

Step 1: Rewrite the integral so that it is with respect to u.

Step 2: Substitute for x and $\dfrac{dx}{du}$.

Step 3: Integrate with respect to u.

Step 4: Rewrite in terms of x.

$$\int (2x - 1)^8\, dx = \int (2x - 1)^8\, \frac{dx}{du}\, du$$
$$= \int u^8 \times \tfrac{1}{2}\, du$$
$$= \tfrac{1}{2} \int u^8\, du$$
$$= \tfrac{1}{2} \times \tfrac{1}{9} u^9 + c$$
$$= \tfrac{1}{18}(2x - 1)^9 + c$$

Side working:
$u = 2x - 1$
$\dfrac{du}{dx} = 2 \Rightarrow \dfrac{dx}{du} = \dfrac{1}{2}$

Recall:
$\dfrac{dx}{du} = \dfrac{1}{\frac{du}{dx}}$

Tip:
You could find $\dfrac{dx}{du}$ by making x the subject, where
$x = \dfrac{u + 1}{2} = \tfrac{1}{2} u + \tfrac{1}{2}$

Example 3.19 Using the substitution $u = 3 - 2x$, find $\int e^{3 - 2x}\, dx$.

Step 1: Rewrite the integral so that it is with respect to u.

Step 2: Substitute for x and $\dfrac{dx}{du}$.

Step 3: Integrate with respect to u.

Step 4: Rewrite in terms of x.

$$\int e^{3 - 2x}\, dx = \int e^{3 - 2x}\, \frac{dx}{du}\, du$$
$$= \int e^u \times \left(-\tfrac{1}{2}\right) du$$
$$= -\tfrac{1}{2} \int e^u\, du$$
$$= -\tfrac{1}{2} e^u + c$$
$$= -\tfrac{1}{2} e^{3 - 2x} + c$$

Side working:
$u = 3 - 2x$
$\dfrac{du}{dx} = -2 \Rightarrow \dfrac{dx}{du} = -\dfrac{1}{2}$

Tip:
Start by doing the side working so that you can get an integral in terms of u.

Example 3.20 Using the substitution $u = 3x + 1$, find $\int \dfrac{1}{3x + 1}\, dx$.

Step 1: Rewrite the integral so that it is with respect to u.

Step 2: Substitute for x and $\dfrac{dx}{du}$.

Step 3: Integrate with respect to u.

Step 4: Rewrite in terms of x.

$$\int \frac{1}{3x + 1}\, dx = \int \frac{1}{3x + 1}\, \frac{dx}{du}\, du$$
$$= \int \frac{1}{u} \times \frac{1}{3}\, du$$
$$= \tfrac{1}{3} \int \frac{1}{u}\, du$$
$$= \tfrac{1}{3} \ln |u| + c$$
$$= \tfrac{1}{3} \ln |3x + 1| + c$$

Side working:
$u = 3x + 1$
$\dfrac{du}{dx} = 3 \Rightarrow \dfrac{dx}{du} = \dfrac{1}{3}$

Each of the above three examples could have been integrated using standard results, derived using the chain rule.

- $\displaystyle \int (ax + b)^n\, dx = \frac{1}{a(n + 1)}(ax + b)^{n + 1} + c$ ①

- $\displaystyle \int e^{ax + b}\, dx = \frac{1}{a} e^{ax + b} + c$ ②

- $\displaystyle \int \frac{1}{ax + b}\, dx = \frac{1}{a} \ln |ax + b| + c$ ③

Tip:
It could save time in the examination if you know these and can use them confidently.

For example:

$$\int \sqrt{5x - 2}\,dx = \int (5x - 2)^{\frac{1}{2}}\,dx = \frac{1}{5 \times \frac{3}{2}}(5x - 2)^{\frac{3}{2}} + c = \tfrac{2}{15}(5x - 2)^{\frac{3}{2}} + c$$

Tip:
Write in index form, then use result ①.

$$\int e^{4x - 2}\,dx = \tfrac{1}{4} e^{4x - 2} + c$$

Tip:
Use result ②.

$$\int \frac{5}{8x + 3}\,dx = \tfrac{5}{8} \ln|8x + 3| + c$$

Tip:
Use result ③. The 5 in the numerator is a numerical factor that can be taken outside the integral.

There are, however, times when the substitution working must be carried out, as in the following example:

Example 3.21 Using the substitution $u = 2x + 1$, find $\int x \sqrt{2x + 1}\,dx$.

Step 1: Rewrite the integral so that it is with respect to u.

Step 2: Substitute for x and $\frac{dx}{du}$.

Step 3: Integrate with respect to u.

Step 4: Rewrite in terms of x.

$$\int x \sqrt{2x + 1}\,dx = \int x \sqrt{2x + 1}\,\frac{dx}{du}\,du$$
$$= \int \tfrac{1}{2}(u - 1)u^{\frac{1}{2}} \times \tfrac{1}{2}\,du$$
$$= \tfrac{1}{4} \int (u^{\frac{3}{2}} - u^{\frac{1}{2}})\,du$$
$$= \tfrac{1}{4} \times \left(\frac{1}{\frac{5}{2}} u^{\frac{5}{2}} - \frac{1}{\frac{3}{2}} u^{\frac{3}{2}} \right) + c$$
$$= \tfrac{1}{4}(\tfrac{2}{5}(2x + 1)^{\frac{5}{2}} - \tfrac{2}{3}(2x + 1)^{\frac{3}{2}}) + c$$
$$= \tfrac{1}{10}(2x + 1)^{\frac{5}{2}} - \tfrac{1}{6}(2x + 1)^{\frac{3}{2}} + c$$

Side working:
$u = 2x + 1$
$\frac{du}{dx} = 2 \Rightarrow \frac{dx}{du} = \frac{1}{2}$
$x = \frac{u - 1}{2} = \tfrac{1}{2}(u - 1)$

Recall:
$\frac{dx}{du} = \frac{1}{\frac{du}{dx}}$ (section 3.4).

Tip:
If you are asked to give this in factorised format, it is easier to take out factors before substituting back for x.

When performing **definite integration** using a substitution, it is advisable to change the limits to the limits of the new variable. This is illustrated in the following example.

Example 3.22 Show that $\int_1^2 \frac{x}{3x + 1}\,dx = p + q \ln r$ where p, q and r are positive rational numbers to be found.

Step 1: Choose an appropriate substitution.

Step 2: Rewrite the integral so that it is with respect to u.

Step 3: Substitute for x, $\frac{dx}{du}$ and the new limits.

Step 4: Integrate with respect to u.

Step 5: Substitute the limits and evaluate.

Step 6: Express in the required format and state the values of p, q and r.

Let $u = 3x + 1$

$$\int_1^2 \frac{x}{3x + 1}\,dx = \int_{x=1}^{x=2} \frac{x}{3x + 1}\,\frac{dx}{du}\,du$$
$$= \int_{u=4}^{u=7} \frac{\tfrac{1}{3}(u - 1)}{u} \times \frac{1}{3}\,du$$
$$= \tfrac{1}{9} \int_4^7 \left(1 - \frac{1}{u}\right)\,du$$
$$= \tfrac{1}{9}[u - \ln|u|]_4^7$$
$$= \tfrac{1}{9}(7 - \ln 7 - (4 - \ln 4))$$
$$= \tfrac{1}{9}(7 - \ln 7 - 4 + \ln 4)$$
$$= \tfrac{1}{9}(3 + \ln \tfrac{4}{7})$$
$$= \tfrac{1}{3} + \tfrac{1}{9} \ln \tfrac{4}{7}$$

Hence $p = \tfrac{1}{3}$, $q = \tfrac{1}{9}$ and $r = \tfrac{4}{7}$.

Side working:
$u = 3x + 1$
$\frac{du}{dx} = 3 \Rightarrow \frac{dx}{du} = \frac{1}{3}$
$x = \frac{u - 1}{3} = \tfrac{1}{3}(u - 1)$

Limits:

x	1	2
u	4	7

Tip:
Taking numerical factors outside before integrating will make your working simpler.

Using integration to find areas

Recall:
Areas (C2 Section 4.4).

In *Core 1* and *Core 2* you used the fact that to find the area of the region bounded by the curve $y = f(x)$, the x-axis and the lines between $x = a$ and $x = b$, evaluate $\int_a^b y\,dx$.

Recall:
The value will be positive if the region is above the x-axis and negative if it is below the x-axis.

Example 3.23 The diagram shows a sketch of the curve $y = 4 - e^{2x}$. The curve crosses the x-axis at P.

a Find the x-coordinate of P.

b Find the exact value of the area bounded by the curve and the coordinate axes.

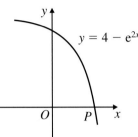

Step 1: Set $y = 0$ and solve.

a At P, $y = 0$

$\Rightarrow \quad 4 - e^{2x} = 0$

$e^{2x} = 4$

$2x = \ln 4$

$x = \tfrac{1}{2}\ln 4 = \ln 2$

The x-coordinate of P is $\ln 2$.

Recall:
$e^a = b \Leftrightarrow a = \ln b$

Recall:
$n\ln a = \ln a^n$

Step 2: Use the area formula.

b $\int_a^b y\,dx = \int_0^{\ln 2} (4 - e^{2x})\,dx$

$= \left[4x - \tfrac{1}{2}e^{2x}\right]_0^{\ln 2}$

$= 4\ln 2 - \tfrac{1}{2}e^{2\ln 2} - (0 - \tfrac{1}{2}e^0)$

$= 4\ln 2 - \tfrac{1}{2}e^{\ln 4} + \tfrac{1}{2}$

$= 4\ln 2 - \tfrac{1}{2}\times 4 + \tfrac{1}{2}$

$= 4\ln 2 - \tfrac{3}{2}$

The area is $4\ln 2 - \tfrac{3}{2}$.

Tip:
$\int e^{kx}\,dx = \tfrac{1}{k}e^{kx} + c$

Tip:
$e^{\ln a} = a$ and $e^0 = 1$

Tip:
Notice the instruction to give the exact value. Your calculator will give an approximate value for $\ln 2$.

3.8 Volumes of revolution

Use definite integration to find a volume of revolution about one of the coordinate axes.

When a region is rotated completely about the x-axis or y-axis, a solid is formed. The volume of this solid is known as the **volume of revolution**.

About the x-axis: $V_x = \pi \int_a^b y^2\,dx$

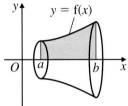

Tip:
You should learn these formulae.

About the y-axis: $V_y = \pi \int_c^d x^2\,dy$

There are several ways of describing the rotation, such as

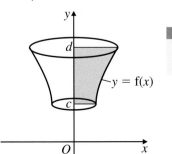

* rotated completely

* rotated through $360°$

* rotated through 2π radians

* rotated through four right angles.

Tip:
Always note carefully which axis the rotation is about.

Example 3.24 The diagram shows the curve $y = \dfrac{1}{2x+1}$.

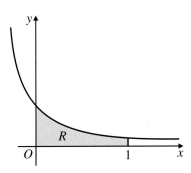

The region R (shown shaded in the diagram) is enclosed by the curve, the axes and the line $x = 1$.

The region R is rotated completely about the x-axis. Find the exact volume of the solid formed.

Step 1: Apply the volume formula.

$$V_x = \pi \int_a^b y^2 \, dx$$

$$= \pi \int_0^1 \left(\frac{1}{2x+1} \right)^2 dx$$

$$= \pi \int_0^1 (2x+1)^{-2} \, dx$$

Step 2: Integrate with respect to x.

$$= \pi \left[\frac{1}{2 \times (-1)} (2x+1)^{-1} \right]_0^1$$

$$= -\tfrac{1}{2}\pi [(2x+1)^{-1}]_0^1$$

Step 3: Substitute the limits and evaluate.

$$= -\tfrac{1}{2}\pi (3^{-1} - 1^{-1})$$

$$= -\tfrac{1}{2}\pi (\tfrac{1}{3} - 1)$$

$$= \tfrac{1}{3}\pi$$

Tip:
A common error is to forget to include π.

Tip:
You must square first, then integrate.

Tip:
Write the expression in index form.

Tip:
$$\int (ax+b)^n = \frac{1}{a(n+1)}(ax+b)^{n+1}$$
If you prefer, substitute $u = 2x + 1$.

Tip:
You are asked for the exact value, so leave your answer in terms of π.

Example 3.25 The diagram shows the curve $y = \ln 2x$ and the lines $y = 1$ and $y = 3$.

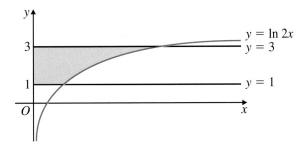

The area bounded by the curve, the y-axis and the lines, shown shaded, is rotated completely about the y-axis. Show that the volume of the solid formed is $\tfrac{1}{8}\pi e^2(e^2 - 1)(e^2 + 1)$.

Step 1: Express x in terms of y.

$$y = \ln 2x \Rightarrow 2x = e^y$$

$$x = \tfrac{1}{2}e^y$$

Recall:
$a = \ln b \Leftrightarrow b = e^a$

Step 2: Apply the volume formula.

$$V_y = \pi \int_c^d x^2 \, dy$$

$$= \pi \int_1^3 \left(\tfrac{1}{2} e^y\right)^2 dy$$

$$= \tfrac{1}{4} \pi \int_1^3 e^{2y} \, dy$$

Step 3: Integrate with respect to y.

$$= \tfrac{1}{4} \pi \left[\tfrac{1}{2} e^{2y}\right]_1^3$$

$$= \tfrac{1}{8} \pi (e^6 - e^2)$$

Step 4: Substitute the limits and simplify to the required form.

$$= \tfrac{1}{8} \pi e^2 (e^4 - 1)$$

$$= \tfrac{1}{8} \pi e^2 (e^2 - 1)(e^2 + 1)$$

Tip:
$(e^a)^2 = e^{2a}$

Tip:
Remember to square the $\tfrac{1}{2}$.

Tip:
Do not round using your calculator.

Tip:
Factorise the difference between two squares.

Example 3.26 The curves $y = x^2$ and $y = \sqrt{x}$ are shown in the sketch.

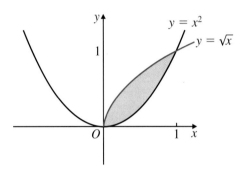

They intersect at $(0, 0)$ and $(1, 1)$. The region enclosed between the curves, shown shaded, is rotated through four right angles about the x-axis. Show that the volume of the solid formed is $\tfrac{3}{10} \pi$.

Note:
This means that it is rotated completely (through 360°).

Step 1: Find the volume when the area 'under' each curve is rotated about the x-axis.

Consider the area under $y = \sqrt{x}$:

$$V_1 = \pi \int_a^b y^2 \, dx$$

$$= \pi \int_0^1 x \, dx$$

$$= \pi \left[\tfrac{1}{2} x^2\right]_0^1$$

$$= \tfrac{1}{2} \pi (1 - 0)$$

$$= \tfrac{1}{2} \pi$$

Consider the area under $y = x^2$:

$$V_2 = \pi \int_a^b y^2 \, dx$$

$$= \pi \int_0^1 x^4 \, dx$$

$$= \pi \left[\tfrac{1}{5} x^5\right]_0^1$$

$$= \tfrac{1}{5} \pi (1 - 0)$$

$$= \tfrac{1}{5} \pi$$

Step 2: Subtract the volumes. The required volume is $V_1 - V_2 = \tfrac{1}{2} \pi - \tfrac{1}{5} \pi = \tfrac{3}{10} \pi$.

Alternatively, you could work out

$$\pi \int_a^b (y_1^2 - y_2^2) \, dx, \quad \text{provided that } y_1 > y_2 \text{ for } a \leqslant x \leqslant b$$

SKILLS CHECK **3D: Integration**

1 Find

a $\displaystyle \int e^{3x+1} \, dx$

b $\displaystyle \int -e^{-u} \, du$

c $\displaystyle \int \frac{1}{e^{2t}} \, dt$

2 Find

a $\displaystyle \int \frac{1}{3x} \, dx$

b $\displaystyle \int \frac{2}{1 + 5x} \, dx$

c $\displaystyle \int \frac{3}{1 - 6x} \, dx$

3 The gradient at the point (x, y) on the curve $y = f(x)$ is given by $e^{3x} - 2x$. The curve goes through $(0, 1)$. Find the equation of the curve.

4 Evaluate

a $\displaystyle\int_0^1 (3x - 1)^5 \, dx$

b $\displaystyle\int_{-\frac{1}{2}}^4 \sqrt{1 + 2x} \, dx$

5 a Show that $\displaystyle\int_5^8 \frac{1}{x - 4} \, dx = 2 \ln 2$

b Evaluate $\displaystyle\int_5^8 \frac{1}{(x - 4)^2} \, dx$

6 Using the substitution $u = 2x + 1$, show each of the following:

a $\displaystyle\int x(2x + 1)^3 \, dx = \frac{1}{80}(8x - 1)(2x + 1)^4 + c$

b $\displaystyle\int_0^1 \frac{x}{(2x + 1)^3} \, dx = \frac{1}{18}$

7 Given that $\displaystyle\int_1^2 \left(\frac{2}{x} - 3x^2\right) dx = a + b \ln 2$, where a and b are integers, find the values of a and b.

8 The diagram shows a sketch of the curve $y = \dfrac{1}{x}$, $x > 0$.

The region R is bounded by the x-axis and the lines $x = 1$ and $x = 2$.

a Find the exact area of R.

b Find the volume of the solid formed when R is rotated completely about the x-axis.

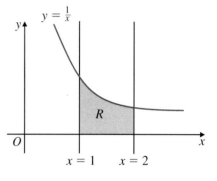

9 The region bounded by the curve $y = e^x$, the coordinate axes and the line $x = 1$ is rotated completely about the x-axis. Find the exact volume of the solid formed.

10 The region bounded by the curve $y = x^3$, the y-axis and the lines $y = 1$ and $y = 8$ is rotated through $360°$ about the y-axis. Find the exact volume of the solid formed.

11 The diagram shows a sketch of the curve $y = \dfrac{1}{\sqrt{x + 1}}$, $x > -1$.

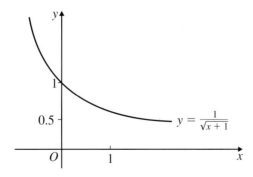

a The region enclosed by the curve, the coordinate axes and the line $x = 1$ is rotated completely about the x-axis. Find the volume of the solid formed.

b The region enclosed by the curve, the y-axis and the lines $y = 0.5$ and $y = 1$ is rotated completely about the y-axis. Find the volume of the solid formed.

12 The diagram shows the curve $y = x^2 + 1$ and the line $y = x + 1$.
The line and the curve intersect at A and B.

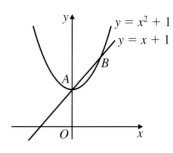

   a Find the coordinates of A and B.

   b The region enclosed between the curve and the line is rotated completely about the x-axis. Find the exact value of the volume of the solid formed.

 13 The diagram shows the curve $y = x(4 - x)$ and the line $y = x$.

The line and the curve intersect at the origin and at P.

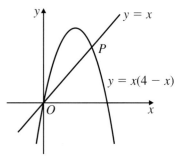

   a Find the coordinates of P.

   b The region enclosed between the curve and the line is rotated completely about the x-axis. Show that the volume of the solid formed is $\frac{108}{5}\pi$.

14 The region enclosed by the curves $y = x^2$ and $y = \sqrt{x}$ is rotated completely about the y–axis (see Example 3.26 for the sketch). Find the exact value of the volume of the solid formed.

SKILLS CHECK **3D EXTRA** is on the CD

Examination practice 3   Differentiation and integration

1 Given that $f(x) = \dfrac{1}{3x + 2}$, find

   i $f'(x)$,   **ii** $\displaystyle\int f(x)\,dx$.

2 Differentiate with respect to x

   i $\dfrac{3}{1 - 2x}$,   **ii** $\ln(4x)$.   [OCR Jan 2004]

3 Find the equation of the tangent to the curve $y = (3x - 4)^5$ at the point for which $x = 2$, giving your answer in the form $y = mx + c$.   [OCR Nov 2003]

4 Find the equation of the tangent to the curve

   $y = (x^2 + 1)^5$

   at the point $(-1, 32)$. Give your answer in the form $y = mx + c$.   [OCR Jan 2002]

5 Find the x-coordinates of the stationary points of $y = x^3\,e^{-kx}$, where k is a positive constant.   [OCR June 1996]

 6 Given that $y = e^x \ln x$, find $\dfrac{d^2y}{dx^2}$.   [OCR June 2000]

7 Find the equation of the normal to the curve $y = \dfrac{x + 1}{2x - 3}$ at the point where $x = 2$. Give your answer in the form $ax + by + c = 0$, where a, b and c are integers.

8 Evaluate $\displaystyle\int_0^1 \frac{1}{(2x + 1)^4}\,dx$, giving your answer as a fraction in its simplest form.

9 Find

   i $\displaystyle\int e^{3-5x}\,dx$ **ii** $\displaystyle\int \frac{1}{4-3x}\,dx$

10

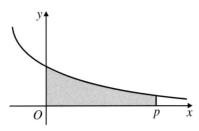

The diagram shows part of the curve with equation

$$y = \frac{9}{(3x+1)^2}.$$

The shaded region lies between the curve and the lines $y = 0$, $x = 0$ and $x = p$, where p is a positive constant. Given that the area of the shaded region is $\frac{7}{3}$, find the exact value of p.   [OCR March 2000]

11 Given that $\displaystyle\int_0^a (2x+a)^3\,dx = 90$, find the exact value of the positive constant a.   [OCR Nov 1999]

 12 Given that $\mathrm{f}(x) = \dfrac{x^2 + 3x + 2}{x + 3}$, where $x < -3$,

   i find $\mathrm{f}'(x)$,

   ii hence solve the equation $\mathrm{f}'(x) = \frac{7}{8}$.

 13 The curve $y = \frac{1}{2}\ln(3x+5)$ crosses the axes at $P(0, p)$ and $Q(q, 0)$.

   a Find the exact values of p and q.

   b Find an equation for the normal to the curve at Q, giving your answer in the form $ay + bx + c = 0$, where a, b and c are integers.

14

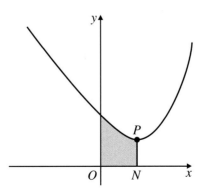

The diagram shows the curve with equation

$$y = 4e^{2x} - 18x.$$

The minimum point of the curve is P. N is the point on the x-axis such that NP is parallel to the y-axis. The region shaded in the diagram is bounded by the curve and the lines $y = 0$, $x = 0$ and NP.

Show that the exact area of the shaded region can be written in the form

$$a - b\left(\ln \tfrac{3}{2}\right)^2,$$

where the values of the constants a and b are to be stated.   [OCR Nov 2003]

15

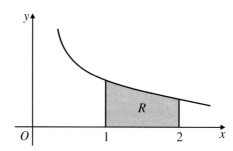

The diagram shows the curve

$$y = \frac{1}{\sqrt{(4x - 1)}}.$$

The region R (shaded in the diagram) is enclosed by part of the curve and by the lines $x = 1$, $x = 2$ and $y = 0$.

i Find the exact area of R.

ii The region R is rotated through four right angles about the x-axis. Find the exact volume of the solid formed.

[OCR June 2002]

16

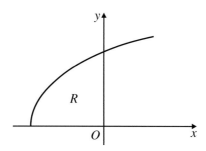

The diagram shows the region R bounded by the axes and part of the curve $y = \sqrt{x + 1}$. Find the exact value of

i the area of R,

ii the volume of the solid formed when R is rotated completely about the y-axis, giving your answer as a multiple of π.

[OCR Nov 1998]

17 A radioactive substance is decaying exponentially. Initially its mass is 480 g. Its mass, M grams, at a time t years after the initial observation is given by

$$M = 480e^{kt},$$

where k is a constant. When $t = 330$, the mass of the substance will be 240 g.

i State the value of the mass when $t = 660$.

ii Determine the value of k.

iii Find the rate at which the mass will be decreasing when $t = 200$. [OCR Jan 2002]

 18 It is given that $y = a^x$, where a is a positive constant.

i Express x in terms of $\ln y$.

ii Find $\dfrac{dx}{dy}$.

iii Hence find $\dfrac{dy}{dx}$ in terms of x.

19 a Given that $y = 5e^{3x} + \ln(2 - 5x)$, find an expression for $\dfrac{dy}{dx}$.

b A water tank is such that, when the depth of water is h cm, the volume, V cm^3, of water in the tank is given by $V = 28h^3$. Water is flowing into the tank at a constant rate of 400 cm^3 per minute. Find the rate at which the depth of water is increasing at the instant when $h = 10$.

[OCR June 2004]

20 The mean annual temperature of the water in a certain large lake is expected to increase due to climatic change. A model giving the mean annual temperature, $\theta°$C, at a time t years after the first observation is

$$\theta = 0.0006t^2 + 0.05t + 4.3.$$

The number of crustaceans in the lake depends on the water temperature. A model giving the number, N, of crustaceans is

$$N = 320e^{0.4\theta}.$$

i According to the models, how many crustaceans were present in the lake when the first observation was made?

ii By first writing down expressions for $\dfrac{d\theta}{dt}$ and $\dfrac{dN}{d\theta}$, find the rate at which the models predict that the number of crustaceans will be increasing when $t = 30$.

[OCR June 2002]

21 a State what is meant by the expression 'exponential growth'. Sketch a graph showing exponential growth.

b A model giving the number, X, of micro-organisms present at time t hours after the start of an experiment is given by

$$X = 275e^{0.018t}.$$

i Find the value of t for which the number of micro-organisms is 2000.

ii Find the rate at which the number of micro-organisms is increasing when $t = 50$.

[OCR Jan 2004]

22 The volume of a sphere is increasing at a rate of 75 cm^3 s^{-1}. Find the rate at which the radius is increasing at the instant when the radius of the sphere is 15 cm. Give your answer correct to 2 significant figures.

[The volume, V, of a sphere of radius r is given by $V = \frac{4}{3}\pi r^3$.]

[OCR Jan 2003]

4 Numerical methods

4.1 Location of roots

Locate approximately a root of an equation, by means of graphical considerations and/or searching for a sign change.

If the function f(x) is continuous in the interval $a \leqslant x \leqslant b$, and f($x$) changes sign in this interval, then f(x) = 0 has a root in the interval $a \leqslant x \leqslant b$.

Example 4.1 It is given that f(x) = sin x − 2x + 1.
Show that f(x) = 0 has a root in the interval $0.8 \leqslant x \leqslant 0.9$.

Step 1: Substitute the boundary values of the interval into f(x) and consider the signs.

f(0.8) = sin 0.8 − 2 × 0.8 + 1 = 0.117.... > 0
f(0.9) = sin 0.9 − 2 × 0.9 + 1 = −0.016... < 0

Change of sign

Step 2: Look for a sign change.

⇒ root of f(x) = 0 lies in the interval $0.8 \leqslant x \leqslant 0.9$

Example 4.2

a Sketch, on the same axes, the graphs of $y = e^x$ and $y = 5 − x$.

Given that f(x) = e^x + x − 5,

b show that the equation f(x) = 0 has only one root,

c show that this root lies in the interval $1 < x < 2$.

Step 1: Sketch the graphs.

a

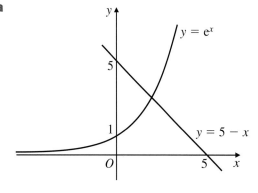

Step 2: Consider the intersection point(s) of the two graphs.

b When the graphs intersect:

$$e^x = 5 − x$$
$$\Rightarrow \quad e^x + x − 5 = 0$$

From the sketch, it can be seen that the graphs of $y = 5 − x$ and $y = e^x$ have only one point of intersection

⇒ there is only one root of e^x + x − 5 = 0

Hence the equation f(x) = 0 has only one root.

Step 3: Substitute the boundary values of the interval into f(x) and consider the signs.

c f(x) = e^x + x − 5
f(1) = e^1 + 1 − 5 = −1.28... < 0
f(2) = e^2 + 2 − 5 = 4.38... > 0

Change of sign ⇒ root of f(x) = 0 in the interval $1 < x < 2$.

4.2 Approximate solutions of equations

Understand the idea of, and use the notation for, a sequence of approximations which converges to a root of an equation; understand how a simple iterative formula of the form $x_{n+1} = F(x_n)$ relates to the equation being solved, and use a given iteration, or an iteration based on a given rearrangement of an equation, to determine a root to a prescribed degree of accuracy.

An approximate solution of an equation can sometimes be found by rearranging the equation into the form $x_{n+1} = F(x_n)$ and then using this iterative formula, with an appropriate starting value, to find subsequent approximations.

Note:
An equation of the form $x_{n+1} = F(x_n)$ is also called a recurrence relation.

If $x_{n+1} = F(x_n)$ converges to a root α, then α is a root of the equation $x = F(x)$.

Note:
A converging sequence approaches a limit.

This method fails if the sequence of values does not converge.

The accuracy of an approximate solution can be tested by using the change of sign method. For example, if a solution of the equation $f(x) = 0$ is 4.327 to three decimal places, then $f(4.3265)$ and $f(4.3275)$ will have different signs.

Example 4.3 The sequence defined by the iterative formula
$$x_{n+1} = \sqrt[3]{7 - 3x_n},$$
with $x_1 = 1.4$, converges to α.

a Use the iterative formula to find α, correct to three decimal places. You should show the result of each iteration.

b State an equation of which α is a root.

Step 1: Apply the iterative formula.

a $x_1 = 1.4$

$x_2 = \sqrt[3]{7 - 3(1.4)} = 1.4094\ldots$

$x_3 = \sqrt[3]{7 - 3(1.4094\ldots)} = 1.4046\ldots$

$x_4 = \sqrt[3]{7 - 3(1.4046\ldots)} = 1.4070\ldots$

$x_5 = \sqrt[3]{7 - 3(1.4070\ldots)} = 1.4058\ldots$

$x_6 = \sqrt[3]{7 - 3(1.4058\ldots)} = 1.4064\ldots$

So $\alpha = 1.406$ (3 d.p.)

Tip:
Use the full calculator display each time.

Tip:
If the values are not converging, check your working.

Calculator note:
If your calculator has the facility of reproducing the last answer, try keying in the following:

| 1.4 | = | $\sqrt[3]{}$ | (| 7 | − | 3 | Ans |) |

The successive terms of the iteration can then be obtained by keying in
| = | | = | | = | etc.

Note:
On graphical calculators, press
| EXE | or | ENTER |

Step 2: Set up the equation $x = F(x)$.

b Since $x_{n+1} = F(x_n)$ converges to a root α, then α is a root of the equation $x = F(x)$.

So $\qquad x = \sqrt[3]{7 - 3x}$

$\qquad\qquad x^3 = 7 - 3x$

$\qquad x^3 + 3x - 7 = 0$

Hence α is a root of the cubic equation $x^3 + 3x - 7 = 0$.

Example 4.4 It is given that $f(x) = \ln x + x + 4$, $x > 0$.

    a Show that $f(x) = 0$ has a root in the interval $(0.01, 0.02)$.

    b Show that $f(x) = 0$ can be rearranged in the form $x = e^{-(x+4)}$.

    c Use an iterative formula based on the equation in part **b**, with $x_1 = 0.01$ to find the values of x_2, x_3 and x_4, giving the value of x_4 to four decimal places.

    d By considering the change of sign of $f(x)$ in a suitable interval, show that your value for x_4 gives an accurate estimate, correct to four decimal places, of the root of $f(x) = 0$.

Tip:
$(0.01, 0.02)$ means $0.01 < x < 0.02$.

Tip:
Even if you can't answer **b**, you could still do parts **c** and **d** as they are not dependent on previous answers.

Step 1: Substitute the boundary values of the interval into $f(x)$ and consider the signs.

a $f(x) = \ln x + x + 4$

     $f(0.01) = \ln 0.01 + 0.01 + 4 = -0.595\ldots < 0$

     $f(0.02) = \ln 0.02 + 0.02 + 4 = 0.107\ldots > 0$

     Change of sign $\Rightarrow$ root of $f(x) = 0$ in $(0.01, 0.02)$

Tip:
Write down the value you get for your calculation to make it clear whether it is positive or negative.

Step 2: Rearrange the equation to the required format.

b $f(x) = 0$

     $\ln x + x + 4 = 0$

     $\ln x = -x - 4 = -(x + 4)$

     $\Rightarrow$ $x = e^{-(x+4)}$

Recall:
$\ln x = y \Leftrightarrow x = e^y$ (Section 1.6).

Step 3: State the iterative formula and find the specified iterations.

c $x_{n+1} = e^{-(x_n + 4)}$

     $x_1 = 0.01$

     $x_2 = e^{-(0.01 + 4)} = 0.01813\ldots$

     $x_3 = e^{-(0.01813\ldots + 4)} = 0.01798\ldots$

     $x_4 = e^{-(0.01798\ldots + 4)} = 0.01798\ldots = 0.0180$ (4 d.p.)

Tip:
Write the formula in the form $x_{n+1} = F(x_n)$.

Tip:
Here the full calculator display has been used each time. Since you are asked for the root to 4 d.p. you must work to at least 5 d.p.

Calculator note:
Try keying in the following:

| 0.01 | = | e^x | (−) | (| Ans | + | 4 |) |

| = | = | = | etc. |

Step 4: Consider the sign of $f(x)$ for appropriate values of x.

d $f(x) = \ln x + x + 4$

     $f(0.01795) = \ln(0.01795) + 0.01795 + 4 = -0.00221\ldots < 0$

     $f(0.01805) = \ln(0.01805) + 0.01805 + 4 = 0.00344\ldots > 0$

     Change of sign $\Rightarrow$ root of $f(x) = 0$ is in $(0.01795, 0.01805)$

     $\Rightarrow$ root is 0.0180 (correct to four decimal places)

Tip:
If the root is 0.0180, correct to four decimal places, then it must lie in the interval $(0.01795, 0.01805)$.

Example 4.5 The sequence defined by the iterative formula

$$x_{n+1} = \sqrt{10 - 2x_n},$$

with $x_1 = 2.3$, converges to α.

    a Use the iterative formula for five iterations. You should show the result of each iteration.

    b State the value of α, correct to two decimal places.

    c State an equation which has α as a root and hence find the exact value of α.

Step 1: Apply the iterative formula.

a $x_1 = 2.3$

$x_2 = \sqrt{10 - 2(2.3)} = 2.3237\ldots$

$x_3 = \sqrt{10 - 2(2.3237\ldots)} = 2.3135\ldots$

$x_4 = \sqrt{10 - 2(2.3135\ldots)} = 2.3179\ldots$

$x_5 = \sqrt{10 - 2(2.3179\ldots)} = 2.3160\ldots$

$x_6 = \sqrt{10 - 2(2.3160\ldots)} = 2.3168\ldots$

Step 2: Give an approximation to the required accuracy.

b $\alpha = 2.32$ (2 d.p.)

Step 3: Set up the equation $x = F(x)$.

c Since $x_{n+1} = F(x_n)$ converges to a root α, then α is a root of the equation $x = F(x)$.

So $\qquad x = \sqrt{10 - 2x}$

$\qquad\qquad x^2 = 10 - 2x$

$\quad x^2 + 2x - 10 = 0$

Hence α is a root of the quadratic equation $x^2 + 2x - 10 = 0$.

Step 4: Solve the quadratic equation.

Completing the square:

$(x + 1)^2 - 1 = 10$

$\qquad (x + 1)^2 = 11$

$\qquad\quad x + 1 = \pm\sqrt{11}$

$\qquad\qquad\quad x = -1 \pm\sqrt{11}$

Step 5: Choose the appropriate root.

Since $\alpha > 0$,

$\qquad \alpha = -1 + \sqrt{11}$

Divergence

Sometimes an iterative process does not lead to a root of the equation even if you use a starting value close to that root. The sequence $x_1, x_2, x_3, \ldots$ may diverge.

Example 4.6 **a** Show that the equation $\frac{1}{2}x^3 - 2x + 1 = 0$ has a root in the interval $(1, 2)$.

b Use the iterative formula $x_{n+1} = \dfrac{x^3 + 2}{4}$ with $x_1 = 1.75$ to find x_2, x_3, x_4, x_5 and x_6, giving your answers to three decimal places.

c Comment on your sequence of values.

Step 1: Substitute the boundary values of the interval into $f(x)$ and consider the signs.

a Let $f(x) = \frac{1}{2}x^3 - 2x + 1$

$f(1) = \frac{1}{2}(1^3) - 2 \times 1 + 1 = -\frac{1}{2} < 0$

$f(2) = \frac{1}{2}(2^3) - 2 \times 2 + 1 = 1 > 0$

Change of sign $\Rightarrow$ root of $f(x) = 0$ in $(1, 2)$.

Step 2: Apply the iterative formula.

b $x_{n+1} = \dfrac{x^3 + 2}{4}$

$x_1 = 1.75$

$x_2 = \dfrac{1.75^3 + 2}{4} = 1.8398\ldots = 1.840$ (3 d.p.)

$x_3 = \dfrac{1.8398\ldots^3 + 2}{4} = 2.0569\ldots = 2.057$ (3 d.p.)

$$x_4 = \frac{2.0569\ldots^3 + 2}{4} = 2.6758\ldots = 2.676 \text{ (3 d.p.)}$$

$$x_5 = \frac{2.6758\ldots^3 + 2}{4} = 5.2899\ldots = 5.290 \text{ (3 d.p.)}$$

$$x_6 = \frac{5.2899\ldots^3 + 2}{4} = 37.5070\ldots = 37.507 \text{ (3 d.p.)}$$

Calculator note:
Try keying in the following:

| 1.75 | = | (| Ans | x^3 | + | 2 |) | ÷ | 4 |

| = | = | = | etc.

Tip:
If you are using a calculator to generate your iterations be careful with brackets and make sure you write down all your intermediate results.

Step 3: Comment on the sequence.

c The sequence diverges. The given iteration does not converge to a root of $\frac{1}{2}x^3 - 2x + 1 = 0$.

Tip:
To find the root in the interval $(1, 2)$, you would need a different iteration formula. Try $x = \sqrt[3]{4x - 2}$.

SKILLS CHECK **4A: Approximate roots of equations**

1 Show that the equation f(x) = 0 has a root in the given interval.

a f(x) = $\sqrt[3]{x} + x - 7$ $\qquad$ $5 < x < 6$

b f(x) = $\cos 2x + x$ $\qquad$ $-1 < x < 0$

c f(x) = $\ln(x - 4) + \sqrt{x}$ $\qquad$ $4.1 \leqslant x \leqslant 4.2$

d f(x) = $\tan x - e^x$ $\qquad$ $-4 < x < -3$

e f(x) = $\dfrac{1}{x} + 1 - x^3$ $\qquad$ $1.2 < x < 1.3$

2 a On the same axes, sketch the graphs of $y = \ln x$ and $y = x^2 - 4$.

b Hence write down the number of roots of the equation $\ln x - x^2 + 4 = 0$.

c Show that the equation $\ln x - x^2 + 4 = 0$ has a root in the interval $2 \leqslant x \leqslant 3$.

3 a Using the same axes, sketch the graphs of $y = e^x - 1$ and $y = 2x + 1$.

b Hence show that the equation $e^x - 2x - 2 = 0$ has one negative root and one positive root.

The positive root of the equation $e^x - 2x - 2 = 0$ lies in the interval $n < x < n + 1$, where n is an integer.

c Find the value of n.

4 In each of the following:

a Show that the equation can be rearranged into the given iterative formula.

b Use the value of x_1 to find the values of x_2, x_3, x_4 and x_5, giving the value of x_5 to the stated accuracy.

i $\quad x^2 - \dfrac{2}{x} - 1 = 0$ $\qquad$ $x_{n+1} = \sqrt{\dfrac{2}{x_n} + 1}$ $\qquad$ $x_1 = 1.5$ $\qquad$ 3 decimal places

ii $\quad \cos x - 9x - 4 = 0$ $\qquad$ $x_{n+1} = \frac{1}{9}(\cos x_n - 4)$ $\qquad$ $x_1 = -0.3$ $\qquad$ 5 decimal places

iii $\ln x + 2 - \sqrt{x} = 0$ $\qquad$ $x_{n+1} = e^{\sqrt{x_n} - 2}$ $\qquad$ $x_1 = 0.2$ $\qquad$ 3 decimal places

iv $\tan x - 3x = 0$ $\qquad$ $x_{n+1} = \tan^{-1}(3x_n)$ $\qquad$ $x_1 = 1.32$ $\qquad$ 3 decimal places

v $e^{0.3x} - x - 2 = 0$ $\qquad$ $x_{n+1} = \dfrac{10}{3}\ln(x_n + 2)$ $\qquad$ $x_1 = 7.51$ $\qquad$ 4 significant figures

5 The sequence defined by the iterative formula

$$x_{n+1} = \sqrt[3]{\frac{x_n + 6}{4}}, \; x_1 = 1.2,$$

converges to α.

   a Use the iterative formula to find α, correct to four significant figures.

   b Find a cubic equation of the form $ax^3 + bx + c = 0$ of which α is a root.

6 The sequence defined by the iterative formula

$$x_{n+1} = \ln(2x_n + 5), \; x_1 = 2.25$$

converges to α.

   a Use the iterative formula to find α, correct to four decimal places.

   b State an equation of which α is a root.

 7 $f(x) = \cos x - x^2 + 3$.

   a Show that $f(x) = 0$ has a root in the interval $1 < x < 2$.

   b Using the iterative formula $x_{n+1} = \sqrt{\cos x_n + 3}$ and $x_1 = 1.7$, write down the values of x_2, x_3 and x_4, giving your answer to x_4 to three decimal places.

8 a On the same axes, sketch the curves with equations $y = 2^x$ and $y = x^3 - 7$.

   b Use your sketch to show that the equation $2^x - x^3 + 7 = 0$ has exactly one solution.

   c Show that the equation $2^x - x^3 + 7 = 0$ can be rearranged to $x = \sqrt[3]{2^x + 7}$

   d Use an appropriate iterative formula with $x_1 = 2.3$ to find x_2, x_3 and x_4.

   e Hence write down an approximate solution of the equation $2^x - x^3 + 7 = 0$.

 9 $f(x) = x^2 - 4x - 8$.

   a Show that $f(x) = 0$ has a root α in the interval $5.4 < x < 5.5$.

   b Use the iterative formula $x_{n+1} = \frac{1}{4}x_n^2 - 2$ and $x_1 = 5.5$ to find x_2, x_3, x_4, x_5 and x_6.

     Comment on your sequence of results.

   c Show that $f(x) = 0$ can be rearranged to the equation $x = \sqrt{4x + 8}$.

   d Use the rearrangement in part **c** to state an iterative formula and use it, with $x_1 = 5.4$, to find α correct to two decimal places.

   e Compare this value with the exact root.

SKILLS CHECK **4A EXTRA** is on the CD

4.3 Numerical integration using Simpson's rule

Carry out numerical integration of functions by means of Simpson's rule.

Simpson's rule is a numerical method for finding an approximate value for the area 'under' a curve.

Since the area bounded by a curve $y = f(x)$, the x-axis and the lines $x = a$ and $x = b$ is given by $\int_a^b y \, dx$, Simpson's rule can be used to estimate the value of this integral.

The area is split into n equal width strips, where n is **even**.

For a better approximation, increase the number of strips.

Note:
There is an odd number of y-values (ordinates).

Simpson's rule:

$$\int_a^b y\,dx \approx \tfrac{1}{3}h\{(y_0 + y_n) + 4(y_1 + y_3 + \cdots + y_{n-1})$$
$$+ 2(y_2 + y_4 + \cdots + y_{n-2})\}$$

where $h = \dfrac{b-a}{n}$ and n is even.

In words: $\int_a^b y\,dx \approx \tfrac{1}{3}h\{\text{ends} + 4(\text{odds}) + 2(\text{evens})\}$

Note:
Simpson's rule is given in the formulae booklet.

Example 4.7   Use Simpson's rule, with 6 strips, to find an approximate value for $\int_0^6 \sqrt{1 + x^3}\,dx$.

Step 1: Calculate h, the width of each strip.

There are 6 strips, so $n = 6$.

$a = 0$, $b = 6$, so $h = \dfrac{6 - 0}{6} = 1$

Step 2: Work out the x-values and the appropriate y-values (ordinates).

Tip:
It can be helpful to write out your working in a table. A possible format is shown here.

	End ordinates	Odd ordinates	Even ordinates
$x_0 = 0$	$y_0 = \sqrt{1}$		
$x_1 = 1$		$y_1 = \sqrt{2}$	
$x_2 = 2$			$y_2 = \sqrt{9}$
$x_3 = 3$		$y_3 = \sqrt{28}$	
$x_4 = 4$			$y_4 = \sqrt{65}$
$x_5 = 5$		$y_5 = \sqrt{126}$	
$x_6 = 6$	$y_6 = \sqrt{217}$		

Step 3: Apply Simpson's rule and evaluate.

$\int_0^6 \sqrt{1 + x^3}\,dx$

$\approx \tfrac{1}{3}h\{(y_0 + y_6) + 4(y_1 + y_3 + y_5) + 2(y_2 + y_4)\}$
$= \tfrac{1}{3} \times 1 \times \{(\sqrt{1} + \sqrt{217}) + 4(\sqrt{2} + \sqrt{28} + \sqrt{126}) + 2(\sqrt{9} + \sqrt{65})\}$
$= \tfrac{1}{3} \times 109.578\ldots$
$= 36.526\ldots$
$= 36.5$ (3 s.f.)

Tip:
To avoid rounding errors, evaluate expressions at the final stage of working where possible.

Example 4.8   Use Simpson's rule, with 4 strips, to find an approximate value for $\int_0^1 \sin^{-1} x\,dx$

Step 1: Calculate h, the width of each strip.

There are 4 strips, so $n = 4$.

$a = 0$, $b = 1$, so $h = \dfrac{1 - 0}{4} = 0.25$

Step 2: Work out the x-values and the appropriate y-values (ordinates).

Tip:
You must work in radians when using trig functions, so remember to set your calculator to radians mode.

	End ordinates	Odd ordinates	Even ordinates
$x_0 = 0$	$y_0 = \sin^{-1}0$		
$x_1 = 0.25$		$y_1 = \sin^{-1}0.25$	
$x_2 = 0.5$			$y_2 = \sin^{-1}0.5$
$x_3 = 0.75$		$y_3 = \sin^{-1}0.75$	
$x_4 = 1$	$y_4 = \sin^{-1}1$		

Step 3: Apply Simpson's rule and evaluate.

$$\int_0^1 \sin^{-1} x \, dx$$

$$\approx \tfrac{1}{3} h\{(y_0 + y_4) + 4(y_1 + y_3) + 2(y_2)\}$$

$$= \tfrac{1}{3} \times 0.25 \times \{(\sin^{-1}0 + \sin^{-1}1) + 4(\sin^{-1}0.25 + \sin^{-1}0.75) + 2(\sin^{-1}0.5)\}$$

$$= \tfrac{1}{3} \times 0.25 \times 7.02096\ldots$$

$$= 0.58508\ldots$$

$$= 0.585 \text{ (3 s.f.)}$$

Tip:
Work out the curly bracket first and write down its value. Then multiply by $\tfrac{1}{3}h$.

Example 4.9 The diagram shows the curve $y = \sec x$ and the lines $x = \tfrac{1}{4}\pi$ and $x = -\tfrac{1}{4}\pi$.

The region R bounded by the curve, the lines and the x-axis is rotated completely about the x-axis.

Use Simpson's rule, with four strips, to find an approximate value for the volume of the solid formed.

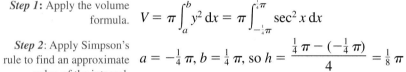

Step 1: Apply the volume formula.

$$V = \pi \int_a^b y^2 \, dx = \pi \int_{-\frac{1}{4}\pi}^{\frac{1}{4}\pi} \sec^2 x \, dx$$

Step 2: Apply Simpson's rule to find an approximate value of the integral.

$$a = -\tfrac{1}{4}\pi, \ b = \tfrac{1}{4}\pi, \text{ so } h = \frac{\tfrac{1}{4}\pi - (-\tfrac{1}{4}\pi)}{4} = \tfrac{1}{8}\pi$$

Tip:
Set your calculator to radians mode.

	End ordinates	**Odd ordinates**	**Even ordinates**
$x_0 = -\tfrac{1}{4}\pi$	$y_0 = \sec^2(-\tfrac{1}{4}\pi)$		
$x_1 = -\tfrac{1}{8}\pi$		$y_1 = \sec^2(-\tfrac{1}{8}\pi)$	
$x_2 = 0$			$y_2 = \sec^2 0$
$x_3 = \tfrac{1}{8}\pi$		$y_3 = \sec^2(\tfrac{1}{8}\pi)$	
$x_4 = \tfrac{1}{4}\pi$	$y_4 = \sec^2(\tfrac{1}{4}\pi)$		

$$\pi \int_{-\frac{1}{4}\pi}^{\frac{1}{4}\pi} \sec^2 x \, dx$$

$$\approx \pi \times \tfrac{1}{3} h\{(y_0 + y_4) + 4(y_1 + y_3) + 2(y_2)\}$$

$$= \pi \times \tfrac{1}{3} \times \tfrac{1}{8}\pi \times \{(\sec^2(-\tfrac{1}{4}\pi) + \sec^2(\tfrac{1}{4}\pi)) + 4(\sec^2(-\tfrac{1}{8}\pi) + \sec^2(\tfrac{1}{8}\pi)) + 2\sec^2(0)\}$$

$$= 6.3217\ldots$$

$$= 6.32 \text{ (3 s.f.)}$$

So the volume of the solid is approximately 6.32.

Note:
You will learn how to integrate $\sec^2 x$ in module C4. The exact answer is $2\pi \ (= 6.283\ldots)$, so Simpson's rule gives a good approximation.

SKILLS CHECK 4B: Numerical integration using Simpson's rule

1 Use Simpson's rule, with 6 strips, to find an approximate value for $\int_1^7 \ln(x^2 + 2) \, dx$.

2 **a** Use Simpson's rule, with 4 strips, to find an approximate value for $\int_0^2 e^{x^2} \, dx$.

   b Explain how to obtain a better approximation using Simpson's rule.

3 **a** Sketch the curve $y = \tan^{-1} x$.

   b Using Simpson's rule with 8 strips, estimate the area of the region bounded by the curve $y = \tan^{-1} x$, the line $x = 1$ and the x-axis.

   c The exact value of the area is $\tfrac{1}{4}\pi - \tfrac{1}{2}\ln 2$. Compare your estimate with this value.

4 **a** Use Simpson's rule, with 6 strips, to find an approximate value for $\int_1^4 \dfrac{1}{1 + x^2} \, dx$.

b The diagram shows the curve $y = \dfrac{1}{\sqrt{1 + x^2}}$.

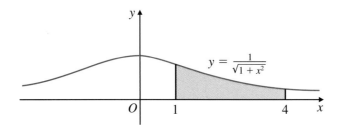

The region bounded by the curve, the x-axis and the lines $x = 1$ and $x = 4$ is rotated completely about the x-axis.

Calculate an estimate of the volume of the solid formed.

SKILLS CHECK **4B EXTRA** is on the CD

Examination practice 4 Numerical methods

1 The sequence given by the iteration formula

$$x_{n + 1} = x_n - 0.15 + 0.05 \cot^2 x_n,$$

with $x_1 = 0.5$, converges to α.

i Find x_2 and x_3 correct to 3 decimal places.

ii State an equation of which α is a root and hence find the exact value of α. [OCR June 2000]

2 Show that the equation $x = \dfrac{1}{3 + \sqrt{x}}$ has a root α between 0 and 1.

By using an iterative formula of the form $x_{n + 1} = F(x_n)$, find α correct to two decimal places. You should show clearly your sequence of approximations.

3 $f(x) = x^3 + \dfrac{2}{x} - 4, \; x \neq 0$

a Show that $f(x) = 0$ has a solution in the interval $0.5 \leqslant x \leqslant 0.6$.

b Show that $f(x) = 0$ can be written in the form $x = \dfrac{2}{4 - x^3}$.

c Using an iterative formula of the form $x_{n + 1} = F(x_n)$, based on the rearrangement in **b**, with $x_1 = 0.52$, calculate the values of x_2, x_3 and x_4, giving your answers to four significant figures.

d Using a change of sign method over a suitable interval, show that the solution of $f(x) = 0$ is 0.5180 correct to four significant figures.

4 $f(x) = 0.1x - \ln (x + 2), \; x > -2$

a Rearrange the equation $f(x) = 0$ into the form $x = e^{ax} + b$, stating the values of the constants a and b.

b Use an iterative formula based on your arrangement in part **a**, with $x_1 = -1.1$ to find x_2, x_3 and x_4.

c Hence write down an approximation to the negative root of the equation $f(x) = 0$.

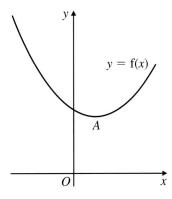

The diagram shows part of the curve with equation $y = f(x)$, where $f(x) = 3x^2 + e^{-x}$. The curve has a minimum at the point A.

a Find $f'(x)$.

b Show that the x-coordinate of A lies in the interval $0.1 < x < 0.2$.

A more accurate estimate of the x-coordinate of A is made using the iterative formula

$x_{n+1} = \frac{1}{6} e^{-x_n}$ with $x_1 = 0.1$.

c Write down the values of x_2, x_3, x_4 and x_5, giving the value of x_5 to three decimal places.

6 The sequence defined by the iterative formula

$$x_{n+1} = \sqrt[3]{(17 - 5x_n)},$$

with $x_1 = 2$, converges to α.

i Use the iterative formula to find α correct to 2 decimal places. You should show the result of each iteration.

ii Find a cubic equation of the form

$$x^3 + cx + d = 0$$

which has α as a root.

[OCR June 2002]

7

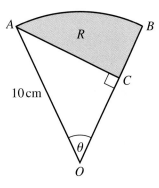

The diagram shows a sector OAB of a circle, centre O and radius 10 cm. Angle AOB is θ radians. The point C lies on OB and is such that AC is perpendicular to OB. The region R (shaded in the diagram) is bounded by the arc AB and by the lines AC and CB. The area of R is 22 cm².

i Show that $\theta = 0.44 + \sin\theta\cos\theta$.

ii Show that θ lies between 0.9 and 1.0.

iii Use an iterative process based on the equation in part **i** to find the value of θ correct to 2 decimal places. You should show the result of each iteration.

[OCR Jan 2002]

8 A curve C has equation

$$y = e^{2x} + x^2 + 4x + 1.$$

i Show that the x-coordinate of any stationary point of C satisfies the equation

$$x = -2 - e^{2x}.$$

ii

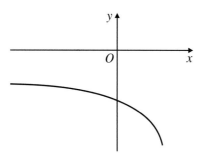

The diagram shows the graph of $y = -2 - e^{2x}$. On a copy of the diagram, draw another graph, the equation of which must be stated, to show that the equation

$$x = -2 - e^{2x}$$

has exactly one root.

iii By using an iteration process based on the equation

$$x = -2 - e^{2x},$$

find, correct to 3 significant figures, the x-coordinate of the stationary point of the curve C. You should show the result of each iteration. [OCR June 2003]

 9 Use Simpson's rule, with 4 strips, to find an approximate value for

$$\int_0^1 x \sin x \, dx$$

giving your answer correct to three decimal places.

10 Use Simpson's rule, with 6 strips, to find an approximate value for

$$\int_0^6 \ln(1 + x^2) \, dx$$

giving your answer correct to four significant figures.

11 a The diagram shows the curve $y = \dfrac{2}{\sqrt{x}}$. The region R, shaded in the diagram, is bounded by the curve and by the lines $x = 1$, $x = 5$ and $y = 0$. The region R is rotated completely about the x-axis. Find the exact volume of the solid formed.

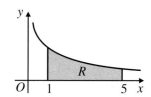

b Use Simpson's rule, with 4 strips, to find an approximate value for

$$\int_1^5 \sqrt{(x^2 + 1)} \, dx,$$

giving your answer correct to 3 decimal places. [OCR June 2005]

Practice exam paper

Answer **all** questions.

Time allowed: 1 hour 30 minutes

A calculator **may** be used in this paper.

1 The curve $y = e^{1-x}$ and the line $y = x + 2$ intersect at the point P.
The x coordinate of P is α.

   i Show that α is a root of the equation $x = 1 - \ln(x + 2)$. *(3 marks)*

   ii Use the iteration formula $x_{n+1} = 1 - \ln(x_n + 2)$ with $x_0 = 0.2$ to find the approximate value of α, correct to three significant figures. *(3 marks)*

2 **i** Solve the equation $2 - |x + 1| = 0$. *(3 marks)*

   ii Sketch the graph of $y = 2 - |x + 1|$, stating the coordinates of the vertex and the intercepts with the axes on the graph. *(3 marks)*

3

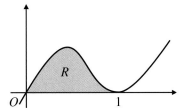

The diagram shows the curve $y = x(x - 1)^2$. The region R is is enclosed between the x-axis and the curve. Use the substitution $u = x - 1$ to evaluate the exact volume of the solid of revolution formed when R is rotated completely about the x-axis. *(7 marks)*

4 The curve $y = \ln x$ intersects the x-axis at $(1, 0)$.

   i Use Simpson's rule, with 4 strips, to find the approximate area enclosed by the curve $y = \ln x$, the x-axis and the line $x = 3$. *(4 marks)*

   ii State the composition of transformations which map the curve $y = \ln x$ onto the curve $y = \frac{1}{2}\ln(x + 1)$. *(2 marks)*

   iii Hence deduce an approximate value for $\displaystyle\int_0^2 \ln\sqrt{(x + 1)}\,dx$. *(1 mark)*

5 **i** By considering the expansion of $\tan(45° + A)$, show that $\tan 75° = 2 + \sqrt{3}$. *(4 marks)*

   ii Find the exact solutions of the equation $\cos 2\theta - 3\cos\theta = 1$, for $0 \leqslant \theta \leqslant 2\pi$. *(4 marks)*

6 **i** Given that $y = xe^{2x}$, express $\dfrac{dy}{dx}$ in terms of x, and show that

$$\frac{d^2y}{dx^2} = 4(1 + x)e^{2x}$$ *(6 marks)*

   ii Hence verify that the curve $y = xe^{2x}$ has a minimum point where $x = -\frac{1}{2}$. *(2 marks)*

7 A small pond which is initially empty, has water poured into it at a constant rate of 3 litres per second. At time t seconds after pouring begins, the depth of water in the pond is x metres and the volume of water is V litres. It is given that $V = 1000\pi x^3$, and that the maximum value of x is 1.

  i Show that when $t = 315$, $x = 0.67$ approximately. *(3 marks)*

  ii Express $\dfrac{dV}{dx}$ in terms of x, and hence find the approximate rate at which the depth of water is increasing when $t = 315$. *(5 marks)*

8 The function f is defined by f: $\theta \mapsto \sin\theta + \sqrt{3}\cos\theta$, $-150° \leqslant \theta \leqslant 30°$.

  i Show that $f(\theta) \equiv R\sin(\theta + 60)$, where R is a constant, and determine the value of R. *(4 marks)*

  ii Hence express $f^{-1}(\theta)$ in terms of θ and evaluate $f^{-1}(1)$. *(4 marks)*

  iii The function g is defined by g: $\theta \mapsto \sin\theta + \sqrt{3}\cos\theta$, $0° \leqslant \theta \leqslant 360°$.
  Solve $g(\theta) = 1$ *(3 marks)*

9 An electrical circuit is constructed from several components including a switch. At time t seconds after the switch is closed, the current in the circuit is x milliamperes. It is given that $t = f(x)$, where

$$f(x) = \ln\left(\frac{10}{10 - x}\right)$$

and the domain of f is $0 \leqslant x < A$.

  i State the greatest value that A can take. *(1 mark)*

  ii Find $f'(x)$. *(3 marks)*

  iii Express $\dfrac{dx}{dt}$ in terms of x, and hence find the value of x when the current is increasing at a rate of 1 milliampere per second. *(3 marks)*

  iv By considering $f^{-1}(x)$, or otherwise, express x in terms of t. *(4 marks)*

Answers

Where graphs are requested in the question, these are provided on the CD.

SKILLS CHECK 1A (page 8)

1 a One-one function   b Not a function
   c One-one function   d Many-one function
2 a f(x) = {2.5, 3, 3.5, 4}   b f(x) = $\{\frac{1}{4}, \frac{1}{3}, \frac{1}{2}, 1\}$
   c $g(x) \in \mathbb{R}$, $g(x) \geqslant 0$ d $f(x) \in \mathbb{R}$, $f(x) \geqslant -2$
3 a See CD; $f(x) \in \mathbb{R}$ b See CD; $g(x) \in \mathbb{R}$, $-1 \leqslant g(x) \leqslant 1$
   c See CD; $f(x) \in \mathbb{R}$, $f(x) > 0$ d See CD; $h(x) \in \mathbb{R}$, $h(x) \geqslant -9$
4 $f(x) \in \mathbb{R}$, $4\sqrt{3} \leqslant f(x) \leqslant 8$
5 a -13 b 4   c 12   d 7   e $3\frac{1}{4}$ f 6
6 a 2^{3x+2} b x c $\frac{3}{x} + 2$ d $\frac{1}{3x+2}$
7 a -4.5 b -4 c $\pm 2\sqrt{2}$ d -3 or -6
8 a No inverse, many-one   b Inverse, one-one
9 a i $f^{-1}: x \mapsto \dfrac{x-5}{2}$, domain $x \in \mathbb{R}$, range $f^{-1}(x) \in \mathbb{R}$
      ii See CD
   b i $f^{-1}: x \mapsto 3 - 4x$, domain $x \in \mathbb{R}$, range $f^{-1}(x) \in \mathbb{R}$
      ii See CD
   c i $f^{-1}: x \mapsto \sqrt{x}$, domain $x \in \mathbb{R}$, $x \geqslant 0$, range $f^{-1}(x) \in \mathbb{R}$, $f^{-1}(x) \geqslant 0$
      ii See CD
   d i $f^{-1}: x \mapsto x^2 + 3$, $x \in \mathbb{R}$, $0 \leqslant x \leqslant 3$, range $f^{-1}(x) \in \mathbb{R}$, $3 \leqslant f^{-1}(x) \leqslant 12$
      ii See CD
10 a $\frac{17}{19}$ b $\frac{1}{2}$ or -1 c $\frac{1}{3}$ or 1
11 a $f^{-1}: x \mapsto \dfrac{2}{3-x}$, $x \neq 3$ b 1 or 2
12 a $\frac{1}{4}$ or 1   b $f^{-1}: x \mapsto \dfrac{5x-1}{4x}$, $x \neq 0$ c -2

SKILLS CHECK 1B (page 17)

1 a Stretch in the y-direction by scale factor $\frac{1}{2}$, translation by -3 units in the y-direction
   b Reflection in the y-axis, stretch in the y-direction by scale factor 0.4
   c Translation by 2 units in the x-direction, stretch in the y-direction by scale factor 5.
2 a See CD; (3, 5), (5, 2)
   b See CD; (0, 1), (1, 0)
   c See CD; $(-2, 0)$, $(0, -3)$, (2, 0)
3 a $y = 3 \tan x - 2$ b $y = -\sin \frac{1}{3} x$
4 a See CD
   b See CD; $(\frac{1}{4}\pi, 0)$, $(\frac{3}{4}\pi, 1)$, $(\frac{5}{4}\pi, 0)$, $(\frac{7}{4}\pi, -1)$
   c $-1 \leqslant g(x) \leqslant 1$
5 a See CD   b See CD   c See CD
      (0, 3), $(\frac{3}{2}, 0)$ $(\frac{1}{2}, 0)$, (0, 1)   (0, 4)
6 a See CD   b $4, 2\frac{2}{5}$ c $2\frac{2}{5} < x < 4$
7 a i See CD; $(-4, 0)$, $(-2, -4)$, (0, 0)
      ii See CD; $(-4, 0)$, $(-2, 4)$, (0, 0)
   b i See CD
      ii See CD
   c i See CD; (0, 1), $(\frac{1}{2}\pi, 0)$, $(\pi, -1)$, $(\frac{3}{2}\pi, 0)$, $(2\pi, 1)$
      ii See CD; (0, 1), $(\frac{1}{2}\pi, 0)$, $(\pi, 1)$, $(\frac{3}{2}\pi, 0)$, $(2\pi, 1)$
8 a See CD   b See CD   c See CD
      (a, 0), (0, a) $(-\frac{a}{2}, 0)$, (0, a)   (0, 2a)
9 a i See CD   b i See CD   c i See CD
      ii $\frac{2}{3}$ or 4   ii 6 or $\frac{4}{3}$ ii 2a or $\frac{2a}{3}$
10 a $-7, 1$ b $-7 < x < 1$ c $x \leqslant -7$, $x \geqslant 1$
11 $\frac{3}{5} < x < 9$

SKILLS CHECK 1C (page 23)

1 $x = 3 \ln 2$
2 $x = \ln 2$
3 $x = \frac{5}{2}$
4 $x = \frac{1}{4}(e^{0.5} - 3)$
5 e^2, e^3

6 a See CD; (1, 0), $x = 0$
   b See CD; $(3 + e^{-2}, 0)$, $x = 3$
7 a i $f^{-1}(x) = \ln \dfrac{x}{3}$ ii See CD; (0, 3), (3, 0)
      iii Range of f: $f(x) \in \mathbb{R}$, $f(x) > 0$, domain of f^{-1}: $x \in \mathbb{R}$, $x > 0$, range of f^{-1}: $f^{-1}(x) \in \mathbb{R}$
   b i $f^{-1}(x) = \frac{1}{2} e^x$ ii See CD $(\frac{1}{2}, 0)$, $(0, \frac{1}{2})$
      iii Range of f: $f(x) \in \mathbb{R}$, domain of f^{-1}: $x \in \mathbb{R}$, range of f^{-1}: $f^{-1}(x) \in \mathbb{R}$, $f^{-1}(x) > 0$
8 a $f^{-1}(x) = e^{4x} - 1$
   b domain of f^{-1}: $x \in \mathbb{R}$, range of f^{-1}: $f^{-1}(x) \in \mathbb{R}$, $f^{-1}(x) > -1$
   c $e^2 - 1$
9 a $f^{-1}(x) = \dfrac{\ln x - 1}{2}$
   b See CD; asymptotes $x = 0$, $y = 0$
   c Range of f: $f(x) \in \mathbb{R}$, $f(x) > 0$, domain of f^{-1}: $x \in \mathbb{R}$, $x > 0$, range of f^{-1}: $f^{-1}(x) \in \mathbb{R}$
10 a £1568.31   b 8 years   c See CD

Exam practice 1 (page 24)

1 i See CD   ii $f^{-1}(x) = -\sqrt{1 - x}$, domain is $x \leqslant 1$
   iii $-\frac{1}{2}$
2 i $f(x) > 2$ ii $f^{-1}(x) = \dfrac{1}{(x-2)^2}$
3 $(x + 2)^2 - 4$, $a = 2$, $b = -4$;
   ii $x \geqslant -4$ iii $f^{-1}(x) = \sqrt{x+4} - 2$,   iv See CD
4 i 28   ii $f^{-1}(x) = (x - 25)^3$,   iii See CD
5 i $f(x) \in \mathbb{R}$, $g(x) \in \mathbb{R}$, $g(x) \geqslant 0$
   iii $\frac{17}{3}$ iv $x = \frac{11}{3}, \frac{13}{3}$
6 $-\frac{3}{14}$
7 i $240 < x < 260$ ii $n = 277, 278, 279, 280$
8 i $f^{-1}(x) = \dfrac{1}{x} - 2$, $x < \frac{1}{2}$ ii $x = 1, 4$
   iii Equation gives $x = -1$ which is not valid since $x > 0$, so no solutions
9 $f^{-1}(x) = e^x - 1$; domain: $x \in \mathbb{R}$, range: $f^{-1}(x) > -1$
   $y = gf(x)$ is a translation of $y = fg(x)$ by $\begin{pmatrix} -1 \\ -1 \end{pmatrix}$
10 i See CD   ii See CD
11 $a = 4$, $b = 2$; $a = -4$, $b = -2$; $x = -1$
12 i $5a$ ii See CD   iii $0, -2a$
13 a $(-0.25, 0)$, (0, 3)   b (0, 2), (0.5, 0)   c $(0, -0.5)$, (1, 0)
14 Translation by $\ln 6$ units in the y-direction; $\ln 6x$, stretch by scale factor $\frac{1}{6}$ in the x-direction
15 i See CD   ii $x = \frac{5}{4} a$
16 i a $4p + 2$ b $3(p + 2)$ c $3p - 1$
   ii $y = e^{-6} x^3$
17 a $x = \ln 4$ b $y = \dfrac{e}{e - 1}$
18 a $\ln 2 - 2$ b $\frac{1}{2} e^3 + 2$
19 i See CD   ii $(-4, 0)$, (0, 8), (0, 2)
20 i See CD for graph: $(0, e - 2)$, $(\ln 2 - 1, 0)$
   ii $f^{-1}(x) = \ln(x + 2) - 1$
   iii Domain: $x \in \mathbb{R}$, $x > -2$; range: $f^{-1}(x) \in \mathbb{R}$
21 i See CD   ii $(2e^{-k}, 0)$
22 i $f(x) > -6$, $g(x) \geqslant 0$ ii $f^{-1}(x) = 2\ln \dfrac{x + 6}{5}$
   iii $x = -2 \ln 5$, $x = 2 \ln 3$

SKILLS CHECK 2A (page 34)

1 a $63°$ b $-42°$ c $70°$ d $102°$
2 a $\frac{1}{4}\pi$ b $-\frac{1}{6}\pi$ c $\frac{1}{2}\pi$ d $\frac{1}{4}\pi$
3 a $23.6°$ (1 d.p.), $156.4°$ (1 d.p.)
   b $150°, 330°$
   c $24.1°$ (1 d.p.), $155.9°$ (1 d.p.), $204.1°$ (1 d.p.), $335.9°$ (1 d.p.)
   d $60°, 120°, 240°, 300°$
4 a 2.38^c (3 s.f.)   b $\frac{3}{2}\pi$
   c 3.39^c (3 s.f.), 6.03^c (3 s.f.)   d $\frac{1}{4}\pi, \frac{3}{4}\pi, \frac{5}{4}\pi, \frac{7}{4}\pi$
5 For proofs see PowerPoint solution on CD
6 $60°, 180°, 300°$
7 $-330°, -210°, 30°, 150°$
8 $\frac{1}{6}\pi, \frac{1}{2}\pi, \frac{5}{6}\pi, \frac{3}{2}\pi$
9 $-\frac{3}{4}\pi, \frac{1}{4}\pi$

10 a Translate $-90°$ in the x-direction, reflect in the x-axis
b See CD
c They are the same curve.

11 a Stretch by scale factor 2 in the y-direction, translate by 1 unit in the y-direction.
b $(\frac{1}{2}\pi, 3)$
c $f(x) \geqslant 3$

12 a Stretch in the x-direction by factor $\frac{1}{2}$, translate by 1 unit in the y-direction.
b $(180, 2)$

SKILLS CHECK 2B (page 38)

1 a $\dfrac{3}{5}$ **b** $\dfrac{5}{13}$ **c** $\dfrac{56}{65}$ **d** $\dfrac{65}{56}$

2 a $\sin(x + 45°) = \sin x \cos 45° + \cos x \sin 45° = \dfrac{1}{\sqrt{2}}\sin x + \dfrac{1}{\sqrt{2}}\cos x$

$= \dfrac{1}{\sqrt{2}}(\sin x + \cos x)$

b $45°, 225°$

3 For proof see PowerPoint solution on CD.

4 a $0°, 180°, 210°, 330°, 360°$ **b** $45°, 90°, 135°, 225°, 270°, 315°$

5 a $0, \frac{2}{3}\pi, \frac{4}{3}\pi, 2\pi$ **b** $0, \frac{4}{3}\pi$

6 a $\text{LHS} = \tan A + \cot A$

$= \dfrac{\sin A}{\cos A} + \dfrac{\cos A}{\sin A}$

$= \dfrac{\sin^2 A + \cos^2 A}{\sin A \cos A}$

$= \dfrac{1}{\frac{1}{2}\sin 2A}$

$= 2 \operatorname{cosec} 2A$

$= \text{RHS}$

b 0.13^c (2 d.p.), 1.44^c (2 d.p.), 3.27^c (2 d.p.), 4.59^c (2 d.p.)

7 a **i** $\cos 2A = 2\cos^2 A - 1$ **ii** $\cos 2A = 1 - 2\sin^2 A$

b $\text{LHS} = \dfrac{2\cos A - \dfrac{1}{\cos A}}{\dfrac{1}{\sin A} - 2\sin A}$

$= \dfrac{\dfrac{2\cos^2 A - 1}{\cos A}}{\dfrac{1 - 2\sin^2 A}{\sin A}}$

$= \dfrac{\cos 2A}{\cos A} \div \dfrac{\cos 2A}{\sin A}$

$= \dfrac{\cos 2A}{\cos A} \times \dfrac{\sin A}{\cos 2A}$

$= \tan A$

$= \text{RHS}$

8 a $\sin(X - Y) = \sin X \cos Y - \cos X \sin Y$

b $\text{LHS} = \dfrac{\sin 4A \cos 2A - \cos 4A \sin 2A}{\sin A}$

$= \dfrac{\sin(4A - 2A)}{\sin A}$

$= \dfrac{\sin 2A}{\sin A}$

$= \dfrac{2\sin A \cos A}{\sin A}$

$= 2\cos A$

$= \text{RHS}$

9 $\frac{1}{2}$

10 For proof see PowerPoint solution on CD.

SKILLS CHECK 2C (page 42)

1 a $\sqrt{5}\cos(x - 26.6°)$ **b** $90°, 323.1°$ (1 d.p.)
2 a $f(\theta) = 5\cos(\theta + 36.9°)$ **b i** 5 **ii** $-36.9°$
3 a $\sqrt{2}\sin(x + \frac{1}{4}\pi)$ **b** $-\frac{1}{12}\pi$
4 a $5\sin(x - 53.1°)$
   b Translate by $53.1°$ in the x-direction and stretch by scale factor 5 in the y-direction
5 a $\alpha = 61.9°$, $k = 8$

Exam practice 2 (page 42)

1 $60°, 109.5°$ (1 d.p.), $250.5°$ (1 d.p.), $300°$
2 $34°$ (nearest °), $135°$, $214°$ (nearest °), $315°$
3 $0, \frac{1}{3}\pi, \frac{5}{3}\pi, 2\pi$
4 $1 + \dfrac{1}{\sqrt{2}}$
5 $\frac{1}{6}(2\sqrt{6} - 1)$
6 See PowerPoint on CD
7 a $25\cos(\theta - 1.287...^c)$ **b** 0.2^c (1 d.p.), 2.3^c (1 d.p.)
8 a $3\sin 2x - 2\cos 2x$
   b $16.8°$ (1 d.p.), $106.8°$ (1 d.p.), $196.8°$ (1 d.p.), $286.8°$ (1 d.p.)
9 i $\dfrac{1}{\cos^2\theta \sin^2\theta}$ **iii** 0.342^c (3 s.f.), 1.23^c (3 s.f.)
10 0.86^c (2 d.p.), 4.00^c (2 d.p.)
11 iii 0.98^c (2 d.p.), 1.89^c (2 d.p.), 4.12^c (2 d.p.), 5.03^c (2 d.p.)
12 i $\sqrt{13}\sin(\theta + \alpha)$, where $\alpha = \tan^{-1}(\frac{2}{3}) = 33.69...°$
   ii $42.2°$ (1 d.p.), $70.2°$ (1 d.p.)

SKILLS CHECK 3A (page 48)

1 a $5e^x$ **b** $\dfrac{3}{x}$ **c** $6x^2 - e^x$ **d** $-\dfrac{2}{x}$

2 a $-6e^{-3x}$ **b** $\dfrac{4}{x}$

   c $3(2x - 6)e^{x^2 - 6x}$ **d** $-\dfrac{8}{5 - 2x}$

3 a $y = \ln 5 + \ln x - \frac{1}{3}\ln(2x + 7)$ **b** $\dfrac{dy}{dx} = \dfrac{1}{x} - \dfrac{2}{3(2x + 7)}$

4 a $\dfrac{dy}{dx} = 20(5x + 6)^3$ **b** $\dfrac{dz}{dx} = \dfrac{1}{\sqrt{2x - 1}}$ **c** $\dfrac{dt}{dx} = -\dfrac{1}{(x + 2)^2}$

5 $y + 6x + 11 = 0$

6 $(2, 4 - 8\ln 2)$

7 a $3e^x - \dfrac{2}{x}$ **b** $y + 2\ln 5 = 3ex - 2x + 2$

   c $2 - \ln 25$

8 a $e^{3x} - 5e^x + 3e^{2x} - 15$ **b** $3e^{3x} - 5e^x + 6e^{2x}$

   c 16

9 -1

10 a $\dfrac{3}{(1 - t)^2}$ **b** $\dfrac{8}{2t + 3}$

11 $f'(x) = e^x - 2x$, $f'(2) = e^2 - 4 = 3.389... > 0$, so function is increasing when $x = 2$

12 a 7389 **b** See CD
   c i $14\,778$/h
     ii Draw the tangent at $t = 1$. It is the gradient of this tangent.

SKILLS CHECK 3B (page 52)

1 a $x^2(x + 4)(5x + 12)$ **b** $2e^{2x}x^3(2 + x)$

   c $\dfrac{x}{2x + 6} + \ln\sqrt{x + 3}$

2 a $\dfrac{3x(x - 6)}{(x - 3)^2}$ **b** $\dfrac{e^{\frac{x}{2}}(x - 6)}{4x^4}$

   c $\dfrac{1 - \ln(x + 1)}{(x + 1)^2}$

3 a $(x^2 + 3)^2(5x - 4)^4(55x^2 - 24x + 75)$

   b $\dfrac{4e^x}{(2e^x + 1)^2}$

   c $\dfrac{x(2 - x)}{e^x}$

4 $-e^{-3}$

5 a $(-1, -2e^{-1})$
   b $(0, 0), (3, 27e^{-3})$

6 b $\dfrac{18}{(x - 3)^3}$

7 $2y + 4x = 5$

8 $\frac{1}{2}$

9 a $\dfrac{dy}{dx} = x \times \frac{1}{2}(1+x)^{-\frac{1}{2}} + (1+x)^{\frac{1}{2}}$

$= \dfrac{x}{2\sqrt{1+x}} + \sqrt{1+x}$

$= \dfrac{x + 2(1+x)}{2\sqrt{1+x}}$

$= \dfrac{2+3x}{2\sqrt{1+x}}$

b $\dfrac{dy}{dx} = \dfrac{(1+2x)1 - 2x}{(1+2x)^2} = \dfrac{1+2x-2x}{(1+2x)^2} = \dfrac{1}{(1+2x)^2}$

10 (e, e^{-1})

11 a 1   **b** $8e^{-2}$

SKILLS CHECK 3C (page 56)

1 $\dfrac{1}{y}$

2 $\dfrac{(y-1)^2}{y^2 - 2y - 1}$

3 $\dfrac{1}{4x^{\frac{3}{4}}}$

4 a $\dfrac{3}{y}$ **b** $6y + 2x = 7$

5 $\dfrac{3}{1 - \ln 3}$

6 a $y = x + 1,\ 2y = x + 4$
  b $(2, 3)$

7 $12y = 13x - 25$

8 a 58 units/s   **b** Decreasing at $78\frac{2}{3}$ unit/s

9 $0.3\pi\,\text{cm}^2$/s

10 0.50 cm/s

SKILLS CHECK 3D (page 62)

1 a $\frac{1}{3}e^{3x+1} + c$ **b** $e^{-u} + c$ **c** $-\frac{1}{2}e^{-2t} + c$

2 a $\frac{1}{3}\ln|x| + c$ **b** $\frac{2}{5}\ln|1+5x| + c$ **c** $-\frac{1}{2}\ln|1-6x| + c$

3 $y = \frac{1}{3}e^{3x} - x^2 + \frac{2}{3}$

4 a 3.5   **b** 9

5 a $[\ln|x - 4|]_5^8 = \ln 4 - 0 = \ln 2^2 = 2\ln 2$
  b 0.75

6 See PowerPoint solution on CD

7 $a = -7,\ b = 2$

8 a $\ln 2$ **b** $\frac{1}{2}\pi$

9 $\frac{1}{2}\pi(e^2 - 1)$

10 $\frac{93}{5}\pi$

11 a $\pi\ln 2$ **b** $\frac{5}{6}\pi$

12 a $A(0, 1), B(1, 2)$ **b** $\frac{7}{15}\pi$

13 a $(3, 3)$

14 a $\frac{3}{10}\pi$

Exam practice 3 (page 64)

1 i $-\dfrac{3}{(3x+2)^2}$ **ii** $\frac{1}{3}\ln|3x+2| + c$

2 i $\dfrac{6}{(1-2x)^2}$ **ii** $\dfrac{1}{x}$

3 $y = 240x - 448$

4 $y = -160x - 128$

5 $x = \dfrac{3}{k},\ x = 0$

6 $e^x\left(-\dfrac{1}{x^2} + \dfrac{2}{x} + \ln x\right)$

7 $x - 5y + 13 = 0$

8 $\frac{13}{81}$

9 i $-\frac{1}{5}e^{3-5x} + c$ **ii** $-\frac{1}{3}\ln|4-3x| + c$

10 $\frac{7}{6}$

11 $\sqrt{3}$

12 i $\dfrac{x^2 + 6x + 7}{(x+3)^2}$ **ii** $x = -7$

13 i $p = \frac{1}{2}\ln 5,\ q = -\frac{4}{3}$ **ii** $9y + 6x + 8 = 0$

14 $a = 2.5,\ b = 9$

15 i $\frac{1}{2}(\sqrt{7} - \sqrt{3})$ **ii** $\frac{1}{4}\pi\ln\frac{7}{3}$

16 i $\frac{2}{3}$ **ii** $\frac{8}{15}\pi$

17 i 120 g   **ii** $-\frac{1}{330}\ln 2$ **iii** 0.662 g/year (3 s.f.)

18 i $x = \dfrac{\ln y}{\ln a}$ **ii** $\dfrac{1}{y\ln a}$ **iii** $a^x \ln a$

19 a $15e^{3x} - \dfrac{5}{2 - 5x}$ **b** 0.048 cm/min

20 i 1787   **ii** 139 per year

21 a The rate of growth is proportional to the quantity present
  b i 110   **iii** 12.2 organisms/hour

22 $0.027\ \text{cm s}^{-1}$

SKILLS CHECK 4A (page 72)

1 a $f(5) = -0.290\ldots, f(6) = 0.817\ldots$
  b $f(-1) = -1.41\ldots, f(0) = 1$
  c $f(4.1) = -0.277\ldots, f(4.2) = 0.439\ldots$
  d $f(-4) = -1.17\ldots, f(-3) = 0.092\ldots$
  e $f(1.2) = 0.105\ldots, f(1.3) = -0.427\ldots$

2 a See CD
  b 2 roots
  c $f(2) = 0.693\ldots, f(3) = -3.90\ldots$

3 a See CD   **c** $n = 1$

4 b i $1.5275\ldots, 1.5196\ldots, 1.5218\ldots, 1.521$
    ii $-0.338295\ldots, -0.339630\ldots, -0.339680\ldots, -0.33968$
    iii $0.2116\ldots, 0.2143\ldots, 0.2150\ldots, 0.215$
    iv $1.3234\ldots, 1.3240\ldots, 1.3241\ldots, 1.324$
    v $7.5078\ldots, 7.5070\ldots, 7.5067\ldots, 7.507$

5 a 1.217 (4 s.f.)   **b** $4x^3 - x - 6 = 0$

6 a 2.2516 (4 d.p.)   **b** $e^x - 2x - 5 = 0$

7 a $f(1) = 2.540\ldots, f(2) = -1.416\ldots$
  b $1.6944\ldots, 1.6960\ldots, 1.696$ (3 d.p.)

8 a See CD
  c $2.28462\ldots, 2.28128\ldots, 2.28056\ldots$
  d 2.28 (2 d.p.)

9 a $f(5.4) = -0.44, f(5.5) = 0.25$
  b $5.5625, 5.7353\ldots, 6.2235\ldots, 7.6831\ldots, 12.7578\ldots$ The sequence is diverging.
  d $x_{n+1} = \sqrt{4x_n + 8},\ 5.46$ (2 d.p.)
  e Exact root (by quadratic formula or completing the square) is $2 + 2\sqrt{3}$.

SKILLS CHECK 4B (page 75)

1 16.566 (3 d.p.)

2 a 17.354 (3 d.p.)
  b Use more strips, ensuring that an even number is chosen.

3 a See Section 2.1
  b 0.4388 (4 d.p.)
  c Very close, so it is a good approximation

4 a 0.5405 (4 d.p.)   **b** 1.6981 (4 d.p.)

Exam practice 4 (page 76)

1 i 0.518 (3 d.p.), 0.522 (3 d.p.)   **ii** $\cot^2 x = 3, x = \frac{1}{6}\pi$

2 $f(0) = -\frac{1}{3}, f(1) = \frac{3}{4}; 0.28$ (2 d.p.)

3 a $f(0.5) = 0.125, f(0.6) = -0.450$ **c** To 4 s.f.: 0.5182, 0.5180, 0.5180
  d $f(0.51795) = 0.000328\ldots, f(0.51805) = -0.000336\ldots$

4 a $a = 0.1, b = -2$
  b $-1.104165\ldots, -1.104538\ldots, -1.104572\ldots$
  c -1.105 (3 d.p.)

5 a $6x - e^{-x}$
  b $f'(0.1) = -0.304\ldots, f'(0.2) = 0.381\ldots$
  c $0.1508\ldots, 0.1433\ldots, 0.1444\ldots, 0.144$ (3 d.p.)

6 i $1.91293\ldots, 1.95179\ldots, 1.93464\ldots, 1.942248\ldots, 1.94$ (2 d.p.)
  ii $x^3 + 5x - 17 = 0$

7 ii If $f(\theta) = \theta - 0.44 - \sin\theta\cos\theta, f(0.9) = -0.0269\ldots < 0,$
    $f(1) = 0.1053\ldots > 0$
  iii $\theta_{n+1} = 0.44 + \sin\theta_n\cos\theta_n; 0.92692\ldots, 0.92010\ldots, 0.92196\ldots, 0.92$ (2 d.p.)

8 ii Draw $y = x$; there is only one point of intersection of the line and the curve

iii $x_{n+1} = -2 - e^{2x_n}$; -2.02 (3 s.f.)

9 0.301 (3 d.p.)

10 12.50 (4 s.f.)

11 a $4\pi \ln 5$ **b** 12.758 (3 d.p.)

Practice exam paper (page 79)

1 ii 0.208 (3 s.f.)

2 i 1, -3

ii

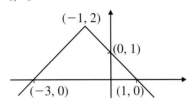

3 $\dfrac{\pi}{105}$

4 i 1.30 (3 s.f.)

ii Translation by -1 unit in the x-direction, stretch by scale factor 0.5 in the y-direction.

iii 0.648 (3 s.f.)

5 ii $\dfrac{2\pi}{3}, \dfrac{4\pi}{3}$

6 i $2xe^{2x} + e^{2x}$

7 ii $3000\pi x^2$, 0.000709 m s^{-1} (3 s.f.)

8 i 2   **ii** $\sin^{-1}\dfrac{\theta}{2} - 60$, $-30°$,   **iii** $90°, 330°$

9 i 10   **ii** $\dfrac{1}{10 - x}$ **iii** $10 - x, 9$ **iv** $10 - 10e^{-t}$

SINGLE USER LICENCE AGREEMENT FOR CORE 3 FOR OCR CD-ROM
IMPORTANT: READ CAREFULLY

WARNING: BY OPENING THE PACKAGE YOU AGREE TO BE BOUND BY THE TERMS OF THE LICENCE AGREEMENT BELOW.

This is a legally binding agreement between You (the user or purchaser) and Pearson Education Limited. By retaining this licence, any software media or accompanying written materials or carrying out any of the permitted activities You agree to be bound by the terms of the licence agreement below.

If You do not agree to these terms then promptly return the entire publication (this licence and all software, written materials, packaging and any other components received with it) with Your sales receipt to Your supplier for a full refund.

YOU ARE PERMITTED TO:

- Use (load into temporary memory or permanent storage) a single copy of the software on only one computer at a time. If this computer is linked to a network then the software may only be used in a manner such that it is not accessible to other machines on the network.

- Transfer the software from one computer to another provided that you only use it on one computer at a time.

- Print a single copy of any PDF file from the CD-ROM for the sole use of the user.

YOU MAY NOT:

- Rent or lease the software or any part of the publication.

- Copy any part of the documentation, except where specifically indicated otherwise.

- Make copies of the software, other than for backup purposes.

- Reverse engineer, decompile or disassemble the software.

- Use the software on more than one computer at a time.

- Install the software on any networked computer in a way that could allow access to it from more than one machine on the network.

- Use the software in any way not specified above without the prior written consent of Pearson Education Limited.

- Print off multiple copies of any PDF file.

ONE COPY ONLY

This licence is for a single user copy of the software

PEARSON EDUCATION LIMITED RESERVES THE RIGHT TO TERMINATE THIS LICENCE BY WRITTEN NOTICE AND TO TAKE ACTION TO RECOVER ANY DAMAGES SUFFERED BY PEARSON EDUCATION LIMITED IF YOU BREACH ANY PROVISION OF THIS AGREEMENT.

Pearson Education Limited and/or its licensors own the software.
You only own the disk on which the software is supplied.

Pearson Education Limited warrants that the diskette or CD-ROM on which the software is supplied is free from defects in materials and workmanship under normal use for ninety (90) days from the date You receive it. This warranty is limited to You and is not transferable. Pearson Education Limited does not warrant that the functions of the software meet Your requirements or that the media is compatible with any computer system on which it is used or that the operation of the software will be unlimited or error free.

You assume responsibility for selecting the software to achieve Your intended results and for the installation of, the use of and the results obtained from the software. The entire liability of Pearson Education Limited and its suppliers and your only remedy shall be replacement free of charge of the components that do not meet this warranty.

This limited warranty is void if any damage has resulted from accident, abuse, misapplication, service or modification by someone other than Pearson Education Limited. In no event shall Pearson Education Limited or its suppliers be liable for any damages whatsoever arising out of installation of the software, even if advised of the possibility of such damages. Pearson Education Limited will not be liable for any loss or damage of any nature suffered by any party as a result of reliance upon or reproduction of or any errors in the content of the publication.

Pearson Education Limited does not limit its liability for death or personal injury caused by its negligence.

This licence agreement shall be governed by and interpreted and construed in accordance with English law.